# SpringerBriefs in Physics

Po-Yuan Chen

# The Application of Biofluid Mechanics

## Boundary Effects on Phoretic Motions of Colloidal Spheres

 Springer

Po-Yuan Chen
Department of Biological Science
  and Technology
China Medical University
Taichung
Taiwan

Additional material to this book can be downloaded from http://extras.springer.com/

ISSN  2191-5423          ISSN 2191-5431   (electronic)
ISBN  978-3-642-44951-2   ISBN 978-3-642-44952-9   (eBook)
DOI 10.1007/978-3-642-44952-9
Springer Heidelberg New York Dordrecht London

Library of Congress Control Number: 2013955056

Printed on acid-free paper

Springer is part of Springer Science+Business Media (www.springer.com)

# Preface

In recent years, the pace of technological innovation is becoming more and more rapid, evolving from the exploration of phenomena from a traditional macroscopic point of view to the research of present microscopic scale biophysical phenomena. Among these researches, the research and development of nanomedicine and nanomaterials are drawing the attention of scientists and scholars, which makes it in a sense approximate to "Getting to know the world from one nut".

The human body is an organism consisting of cells, the size scale of which ranges from several microns to several nanometers. Under such scale, the mobility behavior appears to be very significant, so it is also studied by many experts in biomedical fluid mechanics.

This book aims to discuss various mobility behaviors. The content is divided into two parts: one is the concentration gradient degree as the driving force of diffusion and penetration motions; and the other is temperature gradient-driven thermocapillary and thermophoretic motions. Among this, the diffusiophoresis and penetrate motion are mostly applied in the biomedical field such as drug delivery, purification, as well as the behavior description of immune system, etc.; the thermocapillary and thermophoresis are closely related to semiconductors production and removal of floating impurities. The Appendix contains the comparison and analysis of motion of colloidal particles in the gravitational field situation with the motion action. Eventually, there are relevant computer programs that are summarized into 150 pages. This part is written in FORTRAN language, for scholars to make further applications, and also for the general readers of nonengineering background to appreciate and use as references.

In short, I hope the publication of this book will be an entry for readers interested in motion action.

Po-Yuan Chen

# Contents

# Chapter 1
# Introduction

**Abstract** The transporting behavior of colloidal particles generated by external electric potential, temperature, or gradient of solute concentration in continuous phase is known as mobility. In this study, we concentrate on consideration of mobility of a single spherical colloidal particle parallelling a single infinite plate or two infinite plane walls, and the motion velocity of the particles will be calculated with the boundary collocation method and the reflection method.

## 1.1 Preface

In the case of a low Reynolds number, the delivery behavior of small colloidal particle in a continuous phase is a very critical research topic.

The so-called colloidal particles are small particles with a size in the order of nanometer (nm) or micrometer (μm). Because the total surface area of the unit mass is great, the impact of the physical and chemical properties of the interface between the subject and the surrounding fluid is very huge, so it is worthy further research.

Generally, the driving force of colloidal particles motion includes the diffusion motion of the particles themselves affected by concentration gradient, and the convection motion caused by the sedimentation affected by gravity. Additionally, there are other non-traditional driving forces leading to the colloidal particles motion as well, such as the diffusiophoresis of particles resulted from external solute concentration gradient (Anderson et al. 1982; Anderson and Prieve 1991), and the osmophoresis of vesicle affected by external solute concentration gradient (Gordon 1981; Anderson 1983); the thermophoresis of spherical particles in gaseous medium affected by external temperature gradient (Brock 1962; Keh and Yu 1995) or the thermocapillary motion of droplets affected by temperature gradient (Young et al. 1959; Subramanian 1981), etc. The external solute concentration or temperature field interacts with the particle surface, which generates the particles motion. Anderson (1989) proposed a review article about the detailed introduction

P.-Y. Chen, *The Application of Biofluid Mechanics*, SpringerBriefs in Physics, DOI: 10.1007/978-3-642-44952-9_1, © The Author(s) 2014

of phoretic motions generated by the interaction between the type of field of driving force and the particle surface.

For the motion of a single colloidal particle, there were many previous articles having discussed on it, but there is little discussion on the various motions of particles that are subject to the boundary effect. However, in the real situation of motion behavior, the engagement between colloidal particles and boundary effects lead the phoretic motion apart from behavior of single particle. Thus, it is necessary to explore the boundary effect of the phorectic motions of particles.

This study aims to explore the theory of phoretic motions of single spheric colloidal particle parallel to single or two plane walls of infinite size. The discussion in Chap. 2 is about the diffusiophoresis; Chap. 3 provides an overview of osmophoresis, and the thermocapillary motion is discussed in Chap. 4; thermophoretic motion is investigated in Chap. 5 with boundary collocation method and reflection method; finally, the calculations of these two methods are compared with each other for discussion of boundary effects of overall phoretic particles.

The following sections will initially introduce theories of diffusiophoresis, osmophoresis, thermocapillary motion, and thermophoretic motion and the relevant researches in details through literatures review.

## 1.2  Diffusiophoresis

When the solute molecules in the solution of colloidal particles in which the concentration distribution is uneven due to the physical interactions between the solute and the colloidal particles, the motion of the colloidal particles caused by the concentration gradient of the solute molecules is called diffusiophoresis (Dukhin and Derjaguin 1974); solute molecules could be non-electrolyte (Staffeld and Quinn 1989) or electrolyte (Ebel et al. 1988). When particles are in a non-electrolytic solution with fixed solute concentration gradient, $\nabla C_\infty$, its diffusiophoresis (Anderson et al. 1982) is shown as follow:

$$U^{(0)} = \frac{kT}{\eta} L * K \nabla C_\infty \qquad (1.1)$$

In the formula above, $L*$ is the feature length related to the distance of particle–solute interaction (number level $1 \sim 10$ nm). $K$ means Gibbs adsorption length representing the amount of solute molecules adsorbed on the particle surface ($K$ and $L*$ are defined, respectively, by formula (2.4b) and (2.4c)). $\eta$ is the viscosity of the fluid, $k$ is the Boltzmann's constant, and $T$ is absolute temperature. Formula (1.1) is applied in the case of single particle with any shapes in a borderless solution. However, its derivation must be based on the following two assumptions: first, the radius of local curvature of particle surface should be larger than the thickness (the magnitude of which is equal to $L*$) of interaction layer (diffusion layer) of particle surface and solute molecular; second, the absence of polarization effect of diffusive solute in this interfacial layer, or it should be called the relaxation effect.

During the past few decades, many scholars have focused on the study on motion of polarization of diffusion layer on colloidal particle surface. Anderson and Prieve (1991) analyzed single spherical colloidal particle with a thin diffusion layer with its radius as $a$; in a non-electrolyte solution, in considering the change of solute concentration within the length scale of particle diameter is much smaller than the external concentration of particle center $(a|\nabla C_\infty| << C_\infty)$, and the particle diffusiophoresis velocity could be obtained as below:

$$U_0 = A\nabla C_\infty \tag{1.2a}$$

where

$$A = \frac{kT}{\eta}L^*K(1+\frac{\beta}{a})^{-1} \tag{1.2b}$$

The definition of $\beta$, see (2.3). For strongly adsorbed solutes (such as surfactant), and the relaxation coefficient can be much larger than 1. When $a/\beta$ approaches zero (the adsorption was extremely weak), the polarization effects of diffusive solute in interfacial layer disappears. Formula (1.2) is simplified as formula (1.1). Comparing (1.1) with (1.2), it is recognized that the polarization effect in diffusion layer will slow down the diffusiophoresis velocity of particles. The reason for it is that the solute in interaction layer generates reverse concentration gradient fields along particle surfaces, which could offset external field of driving force (means the imposed solute concentration gradient hereby), thus reducing the phoretic velocity of particles. Even if the change of concentration gradient is more apparent than the radius length of spherical particles, formula (1.2) is still applicable; however, the $\nabla C_\infty$ hereby is the concentration gradient of center of spherical particles, and the particles will not rotate.

If the solute molecules are all adsorbed on the particle surface (means no solute diffusion in adsorption layer), and $L* = 0$ represents there is no diffusiophoresis of particles. For extremely weak adsorption substance, under extreme condition of $\beta/a \rightarrow 0$, there is no polarization effect of diffusive solute in interaction layer, and formula (1.2a, b) can be simplified as formula (1.1).

In real practices of diffusiophoresis, particles do not exist alone, and the fluid around them is often with fixed borders. Hence, the existences of borders of neighboring particles have crucial and usual influences on particles' motion. There is an example for instruction on application of diffusiophoresis in human physiology as follow:

There are several blood cell particles in the human body upon which a variety of driving forces make them flow in the body, so as to supply nutrients, remove waste, resist foreign bacteria violations, among other physiological functions. These driving forces of active transport, muscle tissue from the heart of the extrusion force also have solute concentration gradient of the moving of the blood cell chemotaxis phenomenon. When human is exposed to infection or injury resulting in local inflammation in the endothelial cells and neutrophils adsorbed molecules on the release chemoattractant of adhesion molecule of white blood cells chemoattractant, chemotaxis, or chemotactic factor, produced a concentration gradient, while the neutrophils is in the blood circulation, the diffusion mobility of the small veins to microvascular rear endothelial cells into the inflamed area to resist foreign germ infringement, this phenomenon is called chemical attraction. It is safe to say that the attracting phenomena of the body due to the moving of blood cells caused by the solute concentration gradient is very interesting and important and worth further exploring.

For particles in the extreme conditions of the surface layer (formula (1.1) under the extreme situation) the diffusiophoresis caused by the surrounding fluid due to particle drag regularization velocity field and the case of electrophoresis (Anderson 1989). Therefore, a number of studies were conducted on the electrophoretic particles interaction with boundary effects (Chen and Keh 1999), which can also be used to explain the expansion scattered motion between particles interaction and boundary effects.

When considering the particle diffusion layer around the solute polarization due to the proliferation of diffusiophoresis with electrophoresis loses various particle sizes and some unique factors in the feed mechanism, the diffusiophoresis particle interaction and the behavior of the boundary effect are significantly different from electrophoresis situation. In the past through the boundary collocation method techniques, with thin polarized electric double layer of particles of the vertical plane walls diffusiophoresis. There were several studies on this issue (Keh and Jan 1996). In recent years, a double round column coordinate system of cylindrical particles in the electrolyte solution, thin electrical double layer under the circumstances, the parallel plate vertical to plate two-dimensional diffusion swimming sports. There is also detailed research on this issue (Keh and Hsu 2000).

Chapter 2 provides the study on a rigid spherical particle in any position between a neighboring single plate and two parallel plane walls in the viscous fluid between the plates, to conduct the parallel diffusiophoresis. Its surface characteristics can be divided into solute impenetrable linear in both cases with the solute to be discussed, while interaction between the solute and the plate is assumed to be negligible.

When $\beta/a$ is so small that spherical colloidal particles have diffusiophoresis in closely neighboring plane walls of the solute impermeable, or when $\beta/a$ is so large that the spherical colloidal particles have diffusiophoresis in closely neighboring plane walls of linear distribution of solute concentration, there is a considerably large solute concentration gradient speeding up the motion velocity of spherical colloidal particles. However, the viscosity effects of fluid reduce diffusiophoresis velocity of spherical colloidal particles. The competition of these two forces is apparent increasingly with particles approaching plane walls. Understanding the two forces competing with one another is one of the focuses of this study.

## 1.3 Osmophoretic Motion

In a borderless solvent, if there are differences in the concentration of the solute in it, it will not produce a visible volume of flow. However, when the semipermeable membrane separates the two solutions of different solute concentrations, the membrane allows only solvent to go through, and the solute remain stagnant. It is observed in the solvent that there is a tendency of solution to go over to the lower concentration. This phenomenon is called the osmosis. In fact, dissolved quality is still possible to penetrate a membrane having a selective, but resistance suffered far more from solvent molecules. The reflection coefficient $\sigma$ is used to represent the solute through membrane level. For a semipermeable membrane, $\sigma = 1$; for the non-selective membrane, then $\sigma = 0$. By applying a higher pressure on the solution of a high concentration of the solvent permeate flow; the membrane pressure difference across the $\sigma\Delta\Pi$, which is between the two solution osmophoretic pressures, is stopped. Using the difference multiplied by the coefficient of this membrane excluded $\Pi\sigma$. The low solute concentration (it is considered ideal solution), osmophoretic pressure $\Pi$ solute concentration C meet van't Hoff's principle $\Pi = CRT$. Wherein $R$ and $T$, respectively, represent the gas constant and absolute temperature.

The vesicle particles are a layer of semipermeable membrane coated with the structure of the solution. While it is placed in a non-uniform solute concentration field, one end of vesicle particles has a higher concentration of the solute (i.e., high permeation pressure). This driving force will be such that the solvent internal by vesicle particles penetrating the semipermeable membrane to vesicle particles external. At the end of the low concentration of the solvent, the vesicle particles flow inwardly. At this point, the cell capsule particles like a Micro-Engine and the fluid from one end enters the low concentration region, while the ejection from one end of the high concentration region to vesicle particles will move toward the direction of advancing of the low concentration region. This kind of motion phenomenon caused by the osmophoretic pressure difference is called diffusiophoresis (Gordon 1981; Anderson 1983, 1986). Its motion mechanism can be applied to guide the medicines vesicle particle to certain parts of the body, as well as other related areas of study.

With the rapid development of life sciences and biotechnology in recent years, various issues related to biological cells received wide attention. The extremely thin membrane thickness of a single particle of spherical or ellipsoidal vesicle uniformly dissolved in concentration gradient penetration diffusiophoresis was thoroughly studied by Anderson (1983, 1984). When the particles at a linear distribution of solute concentration of $C_\infty(x)$, considering the particle size of ultrathin vesicle particles of radius $a$ of the thickness of the semipermeable membrane placed solution unboundary diffusiophoresis. In order to simplify this issue, it is assumed that the vesicle particles, internal and external fluids, are Newtonian fluids that are non-compressed and the viscosity values are $\eta$. When the radius of cell capsule particle is small, the convective transport of the inertial force of the fluid and solute can be ignored, which means the Reynolds number and Peclet number are both small, then the relationship of infiltration diffusiophoresis velocity, $U^{(0)}$, and an even solute concentration gradient, $\nabla C_\infty$, can be represented as follow:

$$U^{(0)} = -A\nabla C_\infty \tag{1.3a}$$

where

$$A = aL_pRT(2 + 2\bar{\kappa} + \kappa)^{-1} \tag{1.3b}$$

The dimensionless coefficients $\kappa$ and $\bar{\kappa}$ are defined as follows:

$$\kappa = \frac{a\,L_pRT\,C_0}{D} \tag{1.3c}$$

$$\bar{\kappa} = \frac{a\,L_pRT\,\overline{C}}{\overline{D}} \tag{1.3d}$$

where $L_p$ is the hydraulic coefficient for the fixed membrane and the solvent, and $\overline{D}$ and $D$ represent the vesicle the particles internal and external solute diffusion coefficients, respectively. $C_0$ is vesicle particle inner average solute concentration; and $C_0$ is vesicle particles in the center position of $C_\infty$ value. In Eq. (1.3) derivative mediated process, van't Hoff's Law for description of the relationship between osmophoretic pressure II and solute concentration $C$; if in events that van't Hoff's law does not apply to the range, the RT in (1.3) (1.3c) need to be replaced by $\partial\Pi/\partial C$, and (1.3) wherein $C_0$ and (1.3c) wherein $\overline{C}$ for calculation of differential values. In general, the (1.3a) wherein each parameter in the aqueous solution is typically about $L_p = 10^{-9}\text{m}^2$ s/kg, $|\nabla C_\infty| = 10^5\text{mol/m}^4$ and $\kappa$ (or $\bar{\kappa}$) = 2.5.

From (1.3), it shows that regardless of the size of $C_0$ and $\overline{C}$, vesicle particles are tend low $C_\infty$ direction of moving, while the increase values of the vesicle particles reduce the moving velocity of $\kappa$ or $\bar{\kappa}$ value. In recent years, it was verified experimentally that a radius 10 µm of vesicle particles in the sucrose solution, in the concentration gradient of $10^4$ mol/m$^4$, vesicle particles several microns per second, the velocity of moving (Nardi et al. 1999).

At the real penetration the swimming vesicle particles do not exist alone. They often encounter other vesicle particles and boundary. Using the method of reflections, Anderson (1986) analyzed the $\kappa = \bar{\kappa} = 0$ spherical vesicle particle along the round hole axis and two identical semipermeable membrane of the vesicle particle moving, and the use of two the vesicle particle interactions result of vesicle suspension average penetration of floating liquid diffusiophoresis. The obtained results show that when the vesicle particles are near the boundary, seepage through diffusiophoresis velocity along becomes faster. This phenomenon is due to vesicle particles surrounding solvent flow the direction and the penetration direction of the diffusiophoresis due to the contrary, this phenomenon has been reviewed by Berg and Turner (1990), confirmed through experiments.

In terms of the boundary effect, Keh and Yang (1993a, b) also can double ball coordinates law vesicle semipermeable membrane, moving of the particles is close to a plate. They consider a plate boundary is impermeable in accordance with the plus linear concentration distribution in two cases, the numerical solution of the spherical the vesicle particle penetration diffusiophoresis.

In Chap. 3, boundary collocation method is used in a spherical vesicle particles in a single flat or between any position in two parallel plane walls of viscous fluid, parallel to plane walls penetration diffusiophoresis. The physical characteristics of the plate surface is divided into solute impenetrable linear distribution and solute. The cases are to be investigated. Fluid inertia term is the solute convection term that can be ignored.

Appendix C2 shows: when $\kappa >\ > 1 + \bar{\kappa}$, spherical vesicle impenetrable particles close to the solute plate as a semipermeable membrane diffusiophoresis, or when $\kappa < < 1 + \bar{\kappa}$, the spherical vesicle particle close solute linear distribution of the plate as a semipermeable membrane and diffusiophoresis vesicle particle plate is smaller concentration gradient between degrees so that the fluid out of the vesicle particles effect becomes small and reducing the motion of the particles of spherical vesicle velocity. However, the fluid flow force effects arising out of vesicle particles will be faster than the velocity of spherical vesicle granulocyte the velocity. Understanding the two forces is one of the important parts of the study.

## 1.4 Thermocapillary Motion

When a liquid droplet is suspended in another immiscible fluid, due to the temperature gradient of the effects, causing transported along the droplet surface interfacial tension uneven the action, is called thermocapillary motion. Generally, when the temperature is lower the high interfacial tension, and when the temperature degree is high the low interfacial tension, therefore prompted droplets is moved in the direction of the high temperatures.

In addition to the pure research, the thermocapillary motion is deeply in weightless conditions, thus drawing the researchers' attention. For example, in a weightless state when the material (e.g., high-tech glass or alloy) process, the thermocapillary driven out the unwanted droplets (or bubbles) or by adding the desired droplets (or bubbles).

Young et al. (1959) first proposed the bubble thermocapillary mobility experiment, in theory push operator radius of a spherical droplet suspended in the viscosity $\eta$, there are linear temperature distribution $T_\infty(x)$, thermocapillary motion velocity. If the inertial term is neglected, the fixed temperature gradient $\nabla T_\infty$). The resulting velocity of $U_0$ of the drop thermocapillary motion can be expressed as:

$$U_0 = \frac{2}{(2 + k^*)(2 + 3\eta^*)}(-\frac{\partial \gamma}{\partial T})\frac{a}{\eta}\nabla T_\infty \qquad (1.4)$$

where $\partial_\gamma/\partial T$ is interfacial tension and the change rate with temperature $T$ (the value is about $10^{-4}$ Nm$^{-1}$ K$^{-1}$). Droplets are compared with the internal and external fluid thermal conductivity and viscosity ratio $k^*$ and $\eta^*$, respectively. In formula (1.4), the rest of the physical properties remain unchanged except the interfacial tension. The thermocapillary motion of single bubble can be calculated and assessed by formula (1.4) with the extreme values of $k^* = 0$ and $\eta^* = 0$.

Formula (1.4) is applicable to the thermocapillary motion of single droplet in continuous phase. However, in real practices, the droplet should have borders and presence of several particles rather than exist individually. (Meyyappan et al. 1981, 1983; Anderson 1985; Acrivos et al. 1990; Morton et al. 1990; Satrape 1992; Loewenberg and Davis 1993a, b; Kasumi et al. 2000).

In the past few decades, more and more scholars were about to revise formula (1.4) appropriately with the presence of boundary effects. Meyyappan et al. (1981) and Sadhal (1983) once discussed solving quasi steady state by the double spherical coordinates and single bubble vertical to an infinite plate or fluid interface with thermocapillary motion. Thereafter, Meyyappan and Subramanian (1987), with the same method, calculated the velocity of thermocapillary motion of single bubble in any direction approaching a plane wall. In both cases, when the bubble approaches the boundary, its velocity monotonously decreases.

And then, Barton and Subramanian (1990) and Chen and Keh (1990) further extended the research of Meyyappan et al. (1981) of calculating the plate or fluid interface into a droplet vertical thermostat line with its velocity of thermocapillary motion. In addition, some researchers successfully utilized reflections method (Chen and Keh 1990; Chen 1999) and lubrication method (Loewenberg and Davis 1993b) to obtain the analytical solution of such issues.

The droplet vertical single plate of thermocapillary mobility via experiment (Barton and Subramanian, 1991) was proven to be very consistent with the theoretical estimate its results with the pseudo steady state. Chen et al. (1991) to the border to take point method for solving a single one spherical droplets in a long tube, along the axis thermocapillary motion with the same values of $k^*$ and $\eta^*$, and that, thermocapillary swimming velocity will vary with the liquid dropwise with the cylinder diameter ratio increases monotonously decrease. In the fixed value of $\eta^*$ and the droplet, with the cylinder diameter ratio under regularize, the motion velocity of the droplets will increase gradually with reduced value of $k^*$, which was due to its adiabatic situation internal the tube, when value of $k^*$ reduced, it could increase interfacial tension gradient of droplet. In addition to the foregoing boundary effect on droplet thermocapillary, moving of deformable drop vertical single thermostat plate thermocapillary move also scholars into line (Ascoli and Leal 1990).

In Chap. 4, the detail about the thermocapillary motion solution process of a droplet parallel to single or two plates is discussed. Plate adiabatic or linear temperature degree distribution, can be ignored in the current force of inertia term and the convection heat transfer items deduction. For a relatively low thermal conductivity of droplets near the adiabatic plate thermocapillary motion, or liquid droplets in a high thermal conductivity near linear temperature distribution plate thermocapillary motion dropwise around than in the case of borderless can accumulate a large temperature gradient, resulting in droplets of thermocapillary velocity increases. The viscosity of the fluid force by the plate impact will slow the moving velocity of the droplet. This competing effect of the two forces, the droplets near the plate will increase evidently. Understanding the two competing kinds of forces, is also one of the important parts of the study.

## 1.5  Thermal Motion

The phenomenon of uneven temperature distribution around the aerosol particles of gas (when there is temperature gradient) and aerosol particles move from the high temperature to low temperature, is known as thermophoresis. As early as in 1870, Tyndall has found the thermophoresis phenomenon (Waldmann and Schmitt 1966; Bakanov 1991). He found that when a hot object is placed in a dusty gas, the object will appear around the dust-free area. Kerosene lamps cast of coke, thermophoresis instance: when kerosene causes a temperature combustion, producing a tiny carbon particles, light (high temperature zone), and lampshade (low temperature region) between gradient of carbon particles by thermophoresis adsorbed on the lampshade. In ancient ink sticks the acquisition of raw materials is also thermophoresis, its adsorption in the inner wall of the working room first and then scraped with feathers made of ink sticks.

There are many applications in industry due to thermophoretic deposition phenomenon: thermophoresis activity can remove or collect the particles in the gas leaving the air pure (Batchelor and Shen 1985; Sasse et al. 1994) or as the use of the aerosol particle sampling (Friedlander 1977). Thermophoretic phenomenon can also be used to explore the resulting heat transfer coefficient of the heat exchanger to reduce the causes of institutions modified chemical vapor deposition (Montasssier et al. 1991).

In the processes to manufacture optical fiber, there is an evidence to show that the aerosol particles deposited on the major institutions thermophoretic phenomenon (Simpkins et al. 1979; Weinberg 1982; the Morse et al. 1985). On the other hand, in microelectronic chip manufacturing process, the contamination in the clean room, the particles adhered to the wafer surface is due to thermophoretic (Ye et al. 1991). In terms of nuclear safety, thermophoresis principle can be used to calculate the rate of leakage of nuclear radioactive colloidal particles (Williams 1986; Williams and Loyalka 1991).

The thermophoretic effect can be explained by gas theory (Kennard 1938). As the molecules in the high temperatures have higher kinetic energy than at low temperatures, it prompts particles moving from high temperature to low temperature. Isolated single spherical particles are at a fixed temperature gradient. Thermophoretic velocity of T can be simply expressed as follows:

$$U^{(0)} = -A \, \nabla T_\infty \tag{1.5}$$

where the negative number represents the moving direction of the particles in the direction toward the lower temperature. Thermophoretic mobility A is related to Knudsen number (=l/a; $a$ is the particle radius, and l is the average of the gas molecules free path).

When the Knudsen number is large, the velocity distribution of the gas molecules by small particles is less shadow loud, Maxwell Chapman-Enskog distribution (Waldmann and Schmitt 1966; Whitmore 1981), its thermophoretic movable degree is

$$A = \frac{3\eta}{4(1 + \pi\beta_t/8)\rho_f T_0} \tag{1.5a}$$

where $\eta$ is the viscosity of the fluid, $P_f$ is the fluid density, $T_0$ is the original granulocyte position of the overall gas temperature when the particles do not exist, and $\beta_t$ represents the collision in a heat-reflective surface of aerosol particles the fraction of the number of the total number of gas molecules of the gas molecules, the value of which is usually 0.9 (Friedlander 1977).

When Knudsen number is very small, and Peclet number and Reynolds number are very small and the assumption, and considering the gas—the temperature of the solid surface of the temperature jump, thermal slip, and hydrodynamic slip. Brock (1962) solved a single spherical particle in a fixed temperature gradient $\nabla T_\infty$. thermophoretic velocity is shown as follows:

$$A = \frac{2C_s(k + k_1 C_t l/a)}{(1 + 2C_m l/a)(2k + k_1 + 2kC_t l/a)} \frac{\eta}{\rho_f T_0} \tag{1.5b}$$

In Formula (1.5b), $P_f$, $\eta$ and $k$ represent the density of the gas, viscosity, and thermal conductivity, respectively, compared to the colloidal particle thermal conductivity; $T_0$ is the absolute temperature of the sub-center position of the fluid compared to the aerosol particles that do not exist in the original grain; as for $C_s$, $C_t$, and $C_m$ are dimensionless coefficients, respectively, being thermal slip coefficient, temperature the jump difference and frictional sliding, the values of which are required to be obtained by the experiment. From the experiment reasonable value are obtained: $C_s = 1.17$, $C_t = 2.18$ and $C_m = 1.14$ (Talbot et al. 1980). It is noteworthy that in the ideal gas at constant pressure, $P_f T_0$ in formula (1.5a) is a constant.

In recent years, the calculation of the thermal conductivity of a wide range of Knudsen number and particle thermophoresis of an aerosol particle mobility is a existing research (Li and Davis 1995). According to the calculated results, in critical region Loyalka (1992), the projected linear rigid molecular model Boltzmann equation results match slip part with Brock results (formula (1.5b)).

Thermophoresis in the practical application of the aerosol particles was not alone, it must take into account the inter-particle interaction force boundary existing on the impact of the particle mobility. Chen and Keh (1995) using the border to take points seeking treatment contemplated the thermophoretic under steady state, semi-analytical, and semi-numerical particle vertical single plate of thermal swimming velocity. Thereafter, Chen (2000), reflection method evaluates a single particle near a single plate for thermophoresis movable analytical solution. On the other hand, a single gas thermophoretic aerosol particles in a spherical cavity has also been explored (Keh and Chang 1998; Lu and Lee 2001).

In Chap. 5, I use the border to take point method of rigid spherical aerosol particles near a single plate or between two parallel plane walls, parallel plate thermophoretic motion. The surface can be divided into adiabatic temperature linear distribution of the two situations to explore, and in conservation equations in the temperature of the convection terms with fluid inertia can be ignored.

For a relatively low thermal conductivity of the aerosol particles near adiabatic plate thermophoresis or the aerosol particles on a high thermal conductivity, close to the linear temperature distribution plate thermophoresis around accumulated a large temperature gradient than in the case of no border, resulting in the velocity of the particles of the thermophoresis degree increases. The viscosity of the fluid force will allow the plate to slow down the velocity of the motion of the particles. The two forces of competing effects, particles near the plate will be increasingly evident. Understanding the two forces is also one of the important parts of the study.

# References

Acrivos, A., Jeffrey, D.J., Saville, D.A.: Particle migration in Suspensions by thermocapillary or electrophoretic motion. J. Fluid. Mech. **212**, 95 (1990)

Anderson, J.L.: Movement of a semipermeable vesicle through an osmotic gradient. Phys. Fluids. **26**, 2871 (1983)

Anderson, J.L.: Shape and permeability effects on osmophoresis. PhysicoChem. Hydrodyn. **5**, 205 (1984)

Anderson, J.L.: Droplet interactions in thermocapillary motion. Int. J. Multiph. Flow. **11**, 813 (1985)

Anderson, J.L.: Transport mechanisms of biological colloids. Ann. N. Y. Acad. Sci. (Biochem. Engng IV) **469**, 166 (1986)

Anderson, J.L.: Colloid transport by interfacial forces. Ann. Rev. Fluid Mech. **21**, 61 (1989)

Anderson, J.L., Lowell, M.E., Prieve, D.C.: Motion of a particle generated by chemical gradients. Part 1. Non-electrolytes. J. Fluid Mech. **117**, 107 (1982)

Anderson, J.L., Prieve, D.C.: Diffusiophoresis caused by gradients of stronglyadsorbing solutes. Langmuir **7**, 403 (1991)

Ascoli, E.P., Leal, L.G.: Thermocapillary motion of a deformable drop toward a planar wall. J. Colloid Interface Sci. **138**, 220 (1990)

Bakanov, S.P.: Thermophoresis in gases at small knudsen numbers. Aerosol Sci. Technol. **15**, 77 (1991)

Barton, K.D., Subramanian, R.S.: Thermocapillary migration of a liquid drop normal to a plane surface. J. Colloid Interface Sci. **137**, 170 (1990)

Barton, K.D., Subramanian, R.S.: Migration of liquid drops in a vertical temperature gradient-Interaction effects near a horizontal surface. J. Colloid Interface Sci. **141**, 146 (1991)

Batchelor, G.K., Shen, C.: Thermophoretic deposition of particles in gas flowing over cold surfaces. J. Colloid Interface Sci. **107**, 21 (1985)

Berg, H.C., Turner, L.: Chemotaxis of bacteria in glass capillary arrays. Biophys. J. **58**, 919 (1990)

Brock, J.R.: On the theory of thermal forces acting on aerosol particles. J. Colloid Sci. **17**, 768 (1962)

Chen, J., Dagan, Z., Maldarelli, C.: The axisymmetric thermocapillary motion of a fluid particle in a tube. J. Fluid Mech. **233**, 405 (1991)

Chen, S.B., Keh, H.J.: In Interfacial forces and fields. In: Hsu, J. (ed.). Dekker, New York (1999)

Chen, S.H.: Thermocapillary deposition of a fluid droplet normal to a planar surface. Langmuir **15**, 2674 (1999)

Chen, S.H.: Boundary effects on a thermophoretic sphere in an arbitrary direction of a plane surface. AIChE J. **46**, 2352 (2000)

Chen, S.H., Keh, H.J.: Thermocapillary motion of a fluid droplet normal to a plane surface. J. Colloid Interface Sci. **137**, 550 (1990)

Chen, S.H., Keh, H.J.: Axisymmetric motion of two spherical particles with slip surfaces. J. Colloid Interface Sci. **171**, 63 (1995)

Dukhin, S.S., Derjaguin, B.V.: Electrokinetic pheonmena. In: Matijevic, E.(ed.) J. Colloid Interface Sci., vol. 7. Wiley, New York (1974)

Ebel, J.P., Anderson, J.L., Prieve, D.C.: Diffusiophoresis of latex particles in electrolyte gradients. Langmuir **4**, 396 (1988)

Friedlander, S.K.: Smoke, Dust and Haze. Wiley. New York (1977)

Gordon, L.G.M.: Osmophoresis. J. Phys. Chem. **85**, 1753 (1981)

Kasumi, H., Solomentsev, Y.E., Guelcher, S.A., Anderson, J.L., Sides, P.J.: Thermocapillary flow and aggregation of bubbles on a solid wall. J. Colloid Interface Sci. **232**, 111 (2000)

Keh, H.J., Chang, J.H.: Boundary effects on the creeping-flow and thermophoretic motions of an aerosol particle in a spherical cavity. Chem. Engng. Sci. **53**, 2365 (1998)

Keh, H.J., Hsu, J.H.: Boundary effects on diffusiophoresis of cylindrical particles in nonelectrolyte gradients. J. Colloid Interface Sci. **221**, 210 (2000)

Keh, H.J., Jan, J.S.: Boundary effects on diffusiophoresis and electrophoresis: Motion of a colloidal sphere normal to a plane wall. J. Colloid Interface Sci. **183**, 458 (1996)

Keh, H.J., Yang, F.R.: Boundary effects on osmophoresis: Motion of a vesicle normal to a plane wall. Chem. Engng. Sci. **48**, 609 (1993a)

Keh, H.J., Yang, F.R.: Boundary effects on osmophoresis: Motion of a vesicle in an arbitrary direction with respect to a plane wall. Chem. Engng. Sci. **48**, 3555 (1993b)

Keh, H.J., Yu, J.L.: Migration of aerosol spheres under the combined acting of thermophoretic and gravitational effects. Aerosol Sci. Technol. **22**, 250 (1995)

Kennard, E.H.: Kinetic Theory of Gases. McGraw-Hill, New York (1938)

Li, W., Davis, E.J.: Measurement of the thermophoretic force by electrodynamic levitation: microspheres in air. J. Aerosol Sci. **26**, 1063 (1995)

Loewenberg, M., Davis, R.H.: Near-contact thermocapillary motion of two non-conducting drops. J. Fluid Mech. **256**, 107 (1993a)

Loewenberg, M., Davis, R.H.: Near-contact, thermocapillary migration of a nonconducting, viscous drop normal to a planar interface. J. Colloid Interface Sci. **160**, 265 (1993b)

Loyalka, S.K.: Thermophoretic force on a single particle-i. Numerical solution of the lineralized Boltzmann equation. J. Aerosol Sci. **23**, 291 (1992)

Lu, S.-Y., Lee, C.-T.: Thermophoretic motion of an aerosol particle in a non-concentric pore. J. Aerosol Sci. **32**, 1341 (2001)

Meyyappan, M., Subramanian, R.S.: Thermocapillary migration of a gas bubble in an arbitrary direction with respect to a plane surface. J. Colloid Interface Sci. **115**, 206 (1987)

Meyyappan, M., Wilcox, W.R., Subramanian, R.S.: Thermocapillary migration of a bubble normal to a plane surface. J. Colloid Interface Sci. **83**, 199 (1981)

Meyyappan, M., Wilcox, W.R., Subramanian, R.S.: The slow axisymmetric motion of two bubbles in a thermal gradient. J. Colloid Interface Sci. **94**, 243 (1983)

Montassier, N., Boulaud, D., Renoux, A.: Experimental study of thermophoretic particle deposition in laminar tube flow. J. Aerosol Sci. **22**, 677 (1991)

Morse, T.F., Wang, C.Y., Cipolla, J.W.: Laser-Induced thermophoresis and particle deposition efficiency. J. Heat Transfer **107**, 155 (1985)

Morton, D.S., Subramanian, R.S., Balasubramaniam, R.: The migration of a compound drop due to thermocapillarity. Phys. Fluids A **2**, 2119 (1990)

Nardi, J., Bruinsma, R., Sackmann, E.: Vesicles as osmotic motors. Phys. Rev. Lett. **82**, 5168 (1999)

Sadhal, S.S.: A note on the thermocapillary migration of a bubble normal to a plane surface. J. Colloid Interface Sci. **95**, 283 (1983)

Sasse, A.G.B.M., Nazaroff, W.W., Gadgil, A.J.: Particle filter based on thermophoretic deposition from nature convection flow. Aerosol Sci. Technol. **20**, 227 (1994)

Satrape, J.V.: Interactions and collisions of bubbles in thermocapillary motion. Phys. Fluids A **4**, 1883 (1992)

Simpkins, P.G., Greenberg-Kosinski, S., MacChesney, J.B.: Thermophoresis: The masstransfer mechanism in modified chemical vapor deposition. J. Appl. Phys. **50**, 5676 (1979)

Staffeld, P.O., Quinn, J.A.: Diffusion-induced banding of colloid particles via diffusiophoresis. 2. Non-electrolytes. J. Colloid Interface Sci. **130**, 88 (1989)

Subramanian, R.S.: Slow migration of a gas bubble in a thermal gradient. AIChE J. **27**, 646 (1981)

Talbot, L., Cheng, R.K., Schefer, R.W., Willis, D.R.: Thermophoresis of particles in heated boundary layer. J. Fluid Mech. **101**, 737 (1980)

Waldmann, L., Schmitt, K. H.: Thermophoresis and Diffusiophoresis of Aerosols, Aerosol Science. In: Davies, C.N. (ed.) Academic Press, New York (1966)

Weinberg, M.C.: Thermophoretic efficiency in modified chemical vapor deposition process. J. Am. Ceram. Soc. **65**, 81 (1982)

Whitmore, P.J.: Thermo- and diffusiophoresis for small aerosol particles. J. Aerosol Sci. **12**, 1 (1981)

Williams, M.M.R.: Thermophoretic forces acting on a spheroid. J. Phys. D. **19**, 1631 (1986)

Williams, M.M.R., Loyalka, S.K.: Aerosol Science: Theory and practice, with special applications to the nuclear industry. Pergamon Press, Oxford (1991)

Ye, Y., Pui, D.Y.H., Liu, B.Y.H., Opiolka, S., Blumhorst, S., Fissan, H.: Thermophoretic effect of particle deposition on a free standing semiconductor wafer in a clean room. J. Aerosol Sci. **22**, 63 (1991)

Young, N.O., Goldstein, J.S., Block, M.J.: The motion of bubbles in a vertical temperature gradient. J. Fluid Mech. **6**, 350 (1959)

# Chapter 2
# Diffusiophoresis of Spherical Colloidal Particles Parallel to the Plane Walls

**Abstract** A semi-analytical and semi-numerical calculation is used in single spherical colloidal particles in a nonelectrolyte solution, to calculate the diffusiophoresis velocity without considering the solute convection effect of fluid inertia. The fixed value of the concentration gradient to a parallel plate is the driving force. The boundary conditions for plate can be a solute linear distribution of the two situations that cannot penetrate or solute. When the particle radius is much larger than the thickness of particles and solute interaction layer plate, one part of the boundary effect is from the interaction effect produced by the concentration gradients and the colloidal particles, while the other part is from the viscosity of the fluid. The mobility velocity boundary is used to take point velocity under different polarization parameters and separation parameters to verify the reflection method. Due to the surface characteristics of the particles and the relative distance of the plate from the different boundary conditions on the plate, the plate effect can reduce or increase the motion velocity of particles.

## 2.1 Theoretical Analysis

This chapter concerns the situation of diffusiophoresis parallel to the two flat plane walls of spherical colloidal particles with radius $a$, which is influenced by a nonionic solute concentration gradient.

The distances of central point of spherical particles from the two plane walls are $b$ and $c$, respectively, as shown in Fig. 2.1. In this figure, $(x, y, z)$, $(p, \emptyset, z)$ and $(r, \theta, \varphi)$ represent the Cartesian coordinates with the center of particle as the origin, cylindrical coordinates, and spherical coordinate systems. At infinity, the solute concentration $C_\infty(x)$ displays linear distribution and the concentration gradient $E_\infty e_x$ ($= \nabla C_\infty$, in which $E_\infty$ is positive), while $e_x$, $e_y$, and $e_z$ are the three unit vectors of Cartesian coordinate. With respect to the radius of curvature of the interface and the distance of particles from the boundary, the interaction between fluid and solid interface is negligible. Therefore, the solution around the particles

P.-Y. Chen, *The Application of Biofluid Mechanics*, SpringerBriefs in Physics, 
DOI: 10.1007/978-3-642-44952-9_2, © The Author(s) 2014

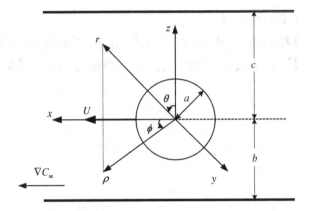

**Fig. 2.1** Geometrical sketch for the diffusiophoresis of a *spherical particle* parallel to two plane walls at an arbitrary position between them

can be divided into inner region and outer region: the inner region is the interface layer of the adjacent layers, and the outer region is the main layer other than the interaction layer. This is estimated based on theoretical calculations to find out with the presence of plane walls, the correction of the velocity of individual particle diffusiophoresis represented by (Eq. 1.2). Before calculating the particle velocity and the fluid velocity, the distribution of the solute concentration of the solution must be found out.

### 2.1.1 Distribution Solute Concentration

Under the circumstance that Reynolds number and Peclet number are quite small and negligible, the mobility state considered can be regarded as a quasisteady state system. In the outer region, the Laplace equation and the Stokes equation can be used, respectively, to represent the conservation equations of fluid concentration and fluid momentum. The distribution of fluid concentration $C$ satisfies

$$\nabla^2 C = 0 \tag{2.1}$$

The boundary conditions required to be met by the outer layer interaction external to the dominating equations are to figure out the internal fluid concentration and the distribution of fluid velocity, and to ensure that in the entire fluid phase, the fluid velocity and fluid concentration are required to have the results of continuity, so that the outcome can be obtained as follows (O'Brien 1983; Anderson and Prieve 1991):

$$r = a: \quad \frac{\partial C}{\partial r} = -\beta [\nabla^2 - \frac{1}{r^2} \frac{\partial}{\partial r} (r^2 \frac{\partial}{\partial r})] C \tag{2.2}$$

In the above formula, $\beta$ is the relaxation coefficient also known as polarization coefficient, and the definition is as follows:

$$\beta = (1 + v\,Pe)K \tag{2.3}$$

where,

$$Pe = \frac{kT}{\eta D} L * KC_0 \tag{2.4a}$$

$$K = \int_0^\infty [\exp(-\Phi(y_n)/kT) - 1]dy_n \tag{2.4b}$$

$$L^* = K^{-1} \int_0^\infty y_n[\exp(-\Phi(y_n)/kT) - 1]dy_n \tag{2.4c}$$

$$v = (L^*K^2)^{-1} \int_0^\infty \{\int_{y_n}^\infty [\exp(-\Phi(y_n')/kT) - 1]dy_n'\}^2 dy_n \tag{2.4d}$$

In (2.4a–d), $\Phi$ represents a solute molecule with the surface of the particles due to the interaction potential arising letter number; $D$ is the fluid diffusion coefficient; $Yn$ is the distance along the particle surface which points to the direction of fluid; $C_0$ is the bulk fluid concentration when particle central location, while the particles do not exist. Fluid concentration away from the particles is not influenced, therefore,

$$z = c,\ -b: \quad \frac{\partial C}{\partial z} = 0 \tag{2.5}$$

$$\rho \to \infty: \quad C = C_\infty = C_0 + E_\infty x \tag{2.6}$$

where, formula (2.5) is the boundary condition that the fluid cannot penetrate the two plane walls, and the relaxation effect of the plate surface has to be ignored. As for the boundary condition of the two flat plane walls which are of linear distribution, formula (2.5) should be changed to

$$z = c,\ -b: \quad C = C_0 + E_\infty x \tag{2.7}$$

Since both governing equation and boundary conditions are linear, fluid concentration of $C$ can be represented as

$$C = C_w + C_p \tag{2.8}$$

where, due to the existence of plate and disturbance caused by the general solution to the double Fourier integral in Cartesian coordinates, as well as the concentration distribution of particles which is undisturbed by the motion, $C_w$ can be expressed as:

$$C_w = C_0 + E_\infty x + E_\infty \int_0^\infty \int_0^\infty (Xe^{\kappa z} + Ye^{-\kappa z}) \sin(\hat{\alpha}x) \cos(\hat{\beta}y) d\hat{\alpha}\, d\hat{\beta} \qquad (2.9)$$

wherein, $X$ and $Y$ are determinable functions, while $\kappa = (\hat{\alpha}^2 + \hat{\beta}^2)^{1/2}$. While $C_p$ satisfies formula (2.1), the spherical coordinate's general solution caused by existence and influence of particles is defined as spherical harmonic function, and is expressed as follows:

$$C_p = E_\infty \sum_{n=1}^\infty R_n r^{-n-1} P_n^1(\mu) \cos \phi \qquad (2.10)$$

wherein, $P_n^1$ is associated Legendre function, $\mu$ represents $\cos\theta$ for the purpose of conciseness, and $R_n$ stands for the unknown coefficient. The concentration distribution of $C$ expressed by formulas (2.8)–(2.10) already satisfies the boundary condition of the infinity formula (2.6).

Apply the fluid concentration distribution $C$ of formulas (2.8)–(2.10) into the boundary condition of formula (2.5) (or (2.7)), and take $x$ and $y$ as the Fourier sine and cosine transforms, then $X$ and $Y$ can be converted into the equation represented by $Rn$. Then, apply its solution into formula (2.9), so the distribution of fluid concentration in Figure $C$ as modified Bessel functions of the second kind is as follows:

$$C = C_0 + E_\infty x + E_\infty \sum_{n=1}^\infty R_n \delta_n^{(1)}(r, \mu) \cos \varphi \qquad (2.11)$$

wherein the detailed definition of the function $\delta_n^{(1)}(r, \mu)$ is in Appendix D of the formula [D1]. Applying Formula (2.11) into the boundary formula (2.2), the following can be obtained:

$$\sum_{n=1}^\infty R_n \left[ \left( \frac{2\beta}{a} - 1 \right) \delta_n^{(2)}(a, \mu) + \beta \delta_n^{(4)}(a, \mu) \right] = \left( 1 - \frac{2\beta}{a} \right)(1 - \mu^2)^{1/2} \qquad (2.12)$$

wherein functions $\delta_n^{(2)}(r, \mu)$ and $\delta_n^{(4)}(r, \mu)$ are defined as [D2] and [D4]. The integrals in the formulas $\delta_n^{(1)}$, $\delta_n^{(2)}$ and $\delta_n^{(4)}$ can be obtained by the number integrals.

In every particle surface, there will be a need for an infinite number of undetermined coefficients $R_n$, so that the boundary conditional (2.12) can be really met. However, using boundary collocation method people can convert infinite series (2.11) into a finite series, and can take a finite number of points on the surface of each particle to satisfy the boundary conditions (O'Brien 1968; Ganatos et al. 1980; Keh and Jan 1996). For each infinite series $\sum_{n=0}^\infty$, the first $M$ item is taken, and then Formula (2.12) involves unknown coefficients $R_n$ which have the number of $M$. Using the $M$ different $\theta_i$ values in each surface of the spherules in formula

(2.11), we can generate $M$ equations which can be used appropriately to solve the undetermined coefficients $R_n$.

### 2.1.2 Distribution of Fluid Velocity

The distribution of the fluid concentration obtained in the previous chapter can be used to further calculate the fluid velocity distribution in this system. Suppose the fluid is incompressible Newtonian fluid, which is creeping flow based on diffusiophoresis, then the outer flow field of diffusion can be conveyed by Stokes equation as follows:

$$\eta \nabla^2 v - \nabla p = 0 \qquad (2.13a)$$

$$\nabla \cdot v = 0 \qquad (2.13b)$$

wherein, $v$ is the velocity distribution of the fluid, while $p$ is its pressure distribution.

Particle surface and the fluid velocity boundary condition at infinity are (Anderson and Prieve 1991):

$$r = a : v = U + a\Omega \times e_r - \frac{kT}{\eta} L^* K (e_\theta e_\theta + e_\varphi e_\varphi) \cdot \nabla C \qquad (2.14)$$

$$z = c, -b : v = 0 \qquad (2.15)$$

$$\rho \to \infty : v = 0 \qquad (2.16)$$

wherein, $e_r$, $e_\theta$ ,and $e_\varphi$ are the unit vectors of the spherical coordinates, and $U = U e_x$ and $\Omega = \Omega e_y$ are the moving velocity and the rotational velocity of the colloidal particles in the diffusion mobility, respectively. Due to the ignorance of the inertial effect, under the symmetrical circumstance (when $b \neq c$), the diffusion velocity of spherical colloidal particles remains parallel to the concentration gradient of the fluid. Since the governing equations and boundary conditions are both linear, the outer region $v$ of the particles can be divided into (Ganatos et al. 1980):

$$v = v_w + v_s \qquad (2.17)$$

wherein, in Formula (2.13a, b), $v_w$ is the Cartesian caused by the influence of the existence of the plane walls.

$$v_w = v_{wx} e_x + v_{wy} e_y + v_{wz} e_z \qquad (2.18)$$

However, $e_x$, $e_y$, and $e_z$, are the three unit vectors of the Cartesian coordinates, respectively, while $v_{wx}$, $v_{wy}$ and $v_{wz}$ are Double Fourier integral.

$$v_{wx} = \int_0^\infty \int_0^\infty D_1(\alpha, \beta, z) \cos(\alpha x) \cos(\beta y) d\alpha \, d\beta \qquad (2.19a)$$

$$v_{wy} = \int_0^\infty \int_0^\infty D_2(\alpha, \beta, z) \sin(\alpha x) \sin(\beta y) d\alpha \, d\beta \qquad (2.19b)$$

$$v_{wz} = v_{wy} = \int_0^\infty \int_0^\infty D_3(\alpha, \beta, z) \sin(\alpha x) \sin(\beta y) d\alpha \, d\beta \qquad (2.19c)$$

In Formula (2.19a–c),

$$D_1 = [X^*(1 + \frac{\alpha^2}{\kappa} z) - X^{**} \frac{\alpha\beta}{\kappa} z - X^{***} \alpha z] e^{\kappa z}$$

$$+ [Y^*(1 - \frac{\alpha^2}{\kappa} z) + Y^{**} \frac{\alpha\beta}{\kappa} z - Y^{***} \alpha z] e^{-\kappa z} \qquad (2.20a)$$

$$D_2 = [-X^* \frac{\alpha\beta}{\kappa} z + X^{**}(1 + \frac{\beta^2}{\kappa} z) + X^{***} \beta z] e^{\kappa z}$$

$$+ [Y^* \frac{\alpha\beta}{\kappa} z + Y^{**}(1 - \frac{\beta^2}{\kappa} z) + Y^{***} \beta z] e^{-\kappa z} \qquad (2.20b)$$

$$D_3 = [X^* \alpha z - X^{**} \beta z + X^{***}(1 - \kappa z)] e^{\kappa z} + [Y^* \alpha z - Y^{**} \beta z + Y^{***}(1 + \kappa z)] e^{-\kappa z} \qquad (2.20c)$$

wherein according to the asterisk mark, $X$ and $Y$ are unknown functions, while $\kappa = (\alpha^2 + \beta^2)^{1/2}$. $V_s$ is the spherical coordinate solution caused by the influence of the presence of the plane walls in formula (2.13a, b) shown as follows:

$$V_s = v_{sx} e_x + v_{sy} e_y + v_{sz} e_z \qquad (2.21)$$

wherein,

$$v_{sx} = \sum_{n=1}^\infty (A_n A_n' + B_n B_n' + C_n C_n') \qquad (2.22a)$$

$$v_{sy} = \sum_{n=1}^\infty (A_n A_n'' + B_n B_n'' + C_n C_n'') \qquad (2.22b)$$

$$v_{sz} = \sum_{n=1}^\infty (A_n A_n''' + B_n B_n''' + C_n C_n''') \qquad (2.22c)$$

For Formula (2.22a–c), according to the marks above, $A_n$, $B_n$, and $C_n$ all involve the Ray built function accompanied by $\mu$ or $\cos\theta$ as variables, the detailed definition of which is provided in Formula (2.6) of the dissertation of Ganatos et al. (1980) (see instructions in Appendix D of this chapter). $A_n$, $B_n$ and $C_n$ are the undetermined coefficients, while formulas (2.17)–(2.22a–c) have been satisfied in the boundary conditions of Formula (2.16) at infinity.

To solve the $X$, $Y$, and, coefficients as well as $A_n$, $B_n$, and $C_n$ from the unknown function, the process is approximated to the process to solve the fluid concentration field. First of all, apply the velocity field v into the boundary conditions of Formula (2.15), and decide the solution of superscripts $X$ and $Y$. Then, apply the general solutions into Formula (2.14), so as to satisfy the boundary conditions of the particle surface, thereby obtaining $A_n$, $B_n$ and $C_n$.

Apply Formulas (2.17)–(2.22a–c) into the boundary conditions of Formula (2.15), and apply Fourier sine and cosine transform to x and y, respectively, then D1, D2, and D3 can be converted into the function of the equations represented by the coefficients, $A_n$, $B_n$, and $C_n$. Then applying this solution back to formula (2.20a–c), Formulas (2.17)–(2.22a–c) can be converted into the modified Bessel functions of the second kind of $A_n$, $B_n$, and $C_n$. The integral types are as follows:

$$v = v_x e_x + v_y e_y + v_z e_z \qquad (2.23)$$

wherein,

$$v_x = \sum_{n=1}^{\infty} [A_n(A_n' + \alpha_n') + B_n(B_n' + \beta_n') + C_n(C_n' + \gamma_n')] \qquad (2.24a)$$

$$v_y = \sum_{n=1}^{\infty} [A_n(A_n'' + \alpha_n'') + B_n(B_n'' + \beta_n'') + C_n(C_n'' + \gamma_n'')] \qquad (2.24b)$$

$$v_z = \sum_{n=1}^{\infty} [A_n(A_n''' + \alpha_n''') + B_n(B_n''' + \beta_n''') + C_n(C_n''' + \gamma_n''')] \qquad (2.24c)$$

In this situation, the marked $\alpha_n$, $\beta_n$ and $\gamma_n$ are integral position functions (they must be obtained by numerical integration), and for their detailed definitions cf the formula [C1] in the chapter of Ganatos et al. (1980) (see the Appendix D of this chapter for instructions).

In order to satisfy the boundary conditions of the particle surface, apply Formulas (2.11) and (2.23) in Formula (2.14), then the outcome is:

$$\sum_{n=1}^{\infty} [A_n(A_n' + \alpha_n') + B_n(B_n' + \beta_n') + C_n(C_n' + \gamma_n')]$$
$$= U + a\Omega\mu - U^{(0)}(H_1\mu\cos\varphi + H_2\sin^2\varphi) \qquad (2.25a)$$

$$\sum_{n=1}^{\infty} [A_n(A_n'' + \alpha_n'') + B_n(B_n'' + \beta_n'') + C_n(C_n'' + \gamma_n'')]$$
$$= -U^{(0)}(H_1\mu\sin\varphi - H_2\cos\varphi\sin\varphi) \qquad (2.25b)$$

$$\sum_{n=1}^{\infty} [A_n(A_n''' + \alpha_n''') + B_n(B_n''' + \beta_n''') + C_n(C_n''' + \gamma_n''')]$$

$$= -a\Omega(1 - \mu^2)^{1/2} \cos\varphi + U^{(0)} H_1 (1 - \mu^2)^{1/2} \tag{2.25c}$$

Therein,

$$H_1 = \mu \cos\varphi + \frac{1}{a} \sum_{n=1}^{\infty} R_n \delta_n^{(3)}(a, \mu) \cos\varphi \tag{2.26a}$$

$$H_2 = 1 + \frac{1}{a(1 - \mu^2)^{1/2}} \sum_{n=1}^{\infty} R_n \delta_n^{(1)}(a, \mu) \tag{2.26b}$$

$U^{(0)} = kTL * KE_\infty/\eta$, while the definitions of equation $\delta_n^{(3)}(r, \mu)$ can be found in Formula [D3]. The item $M$ before coefficient $R_n$ can be obtained by the formulas in the previous chapter.

Observing Formula (2.25a–c) in detail, it can be found that in the spherical surface, when the $r = a$ boundary takes point, all the simultaneous formulas are irrelevant with the selection of the value of $\varphi$. Therefore, Formula (2.25a–c) satisfies $N$ different $\theta_i$ values of the grain surface of each particle ($\theta$ values range between 0 and $\pi$) hence results in $3N$ linear equations, and can solve $3N$ unknown $A_n$, $B_n$ and $C_n$. At the same time, when $N$ is large enough, it can be used to successfully solve the flow distribution.

### 2.1.3 The Deduction of Particle Diffusiophoresis Velocity

The drag force of fluid applied to the spherical particles can be expressed as (Ganatos et al. 1980):

$$F = -8\pi\eta A_1 e_x \tag{2.27a}$$

$$T = -8\pi\eta C_1 e_y \tag{2.27b}$$

From the above equation, it can be known that in Formula (2.24a–c), only low-order coefficients $A_1$ and $C_1$ have contributions to the drag forces and moments applied to spherical particles.

Since the particles can freely suspend in fluid, the particles have received a net force and net torque of zero. Applying this limit to Formula (2.27a–b), the outcome is as follows:

$$A_1 = C_1 = 0 \tag{2.28}$$

Combining Formula (2.28) and the $3N$ linear equations generated by Formula (2.25a–c), the moving velocity of the particles $U$ and the rotational velocity $\Omega$ can be successfully obtained.

### 2.1.4 Calculation Methods of Figures

This chapter explains when the particles move parallel to the two plane walls, and how to find a point on the particle surface to calculate the velocity of the motion of the particles for single particle diffusion. When finding the point on the boundary, I use any meridian plane of the particle (use the $z$-axis as a symmetry axis) to find point on semicircular surface so as to satisfy the boundary conditions. In the first place, I select three points of $\theta_i = 0$, $\pi/2$ and $\pi$, because these points mainly control the particle size in the direction of motion of the projection plane and the distance between the particles and the plane walls, but examining the linear algebraic equations carefully, I found that, if the point positions fall on $\theta_i = 0$, $\pi/2$ or $\pi$, then a set of singular coefficient matrixes will be generated. To avoid this obstacle, four basic points $\theta i = \alpha$, $\pi/2$-$\alpha$, $\pi/2 + \alpha$, and $\pi$-$\alpha$ are selected on the semicircular surface, and the rest of the points are selected along the semicircular arc with $\theta_i = \pi/2$ as the mirror-image pair, and dividing this semicircular arc into the same length. Third in this case, I found that the best value of $\alpha$ is 0.1°. Under this circumstance, all the undetermined coefficients will have satisfactory convergence values. All the numerical calculations in the estimating functions $\alpha_n$, $\beta_n$ and $\gamma_n$ and $\delta_n^{(i)}$ will be obtained by the 80-points approach of Gauss-Laguerre quadrature.

## 2.2 Results and Discussions

Based on this, it will be used to explore the usage of the method of border point in terms of solving a single spherical particle parallel to the two plane walls, and to find out the calculation results of diffusion mobility.

### 2.2.1 Diffusiophoresis of Particle Parallel to One Single Plate

The collocation solutions for the translational and rotational velocities of a spherical particle undergoing diffusiophoresis parallel to a plane wall (with $c$ approaches infinite) for different values of the relaxation parameter and separation parameter $a/b$ are presented in Table 2.1 and 2.2 for the cases of an impermeable wall and a wall with the imposed far-field solute concentration gradient, respectively.

Both the tables use the border method to take point and converge numerical calculations to present effective digits. When the convergence velocity has $a/b$ value, the greater $a/b$, the slower convergence velocity is, and the more the number

**Table 2.1** Diffusiophoretic motion and rotating velocity of *spherical particles* parallel to a single plate that cannot penetrate through the solute

| $a/b$ | $U/U_0$ | | | $-a\Omega/U_0$ | | |
|---|---|---|---|---|---|---|
| | Exact solution[†] | | Asymptotic solution | Exact solution[†] | | Asymptotic solution |
| $\beta/a = 0$ | | | | | | |
| 0.2 | 0.99953 | (0.99953) | 0.99953 | 0.00030 | (0.00030) | 0.00031 |
| 0.4 | 0.99684 | (0.99684) | 0.99688 | 0.00492 | (0.00492) | 0.00540 |
| 0.6 | 0.99172 | (0.99172) | 0.99166 | 0.02669 | (0.02669) | 0.03455 |
| 0.8 | 0.98853 | (0.98853) | 0.98336 | 0.10164 | (0.10164) | 0.15360 |
| 0.9 | 0.99789 | (0.99789) | 0.97635 | 0.20389 | (0.20389) | 0.29818 |
| 0.95 | 1.0223 | (1.02231) | 0.97135 | 0.3189 | (0.31890) | 0.40846 |
| 0.99 | 1.1450 | (1.14536) | 0.96629 | 0.6162 | (0.61832) | 0.52144 |
| 0.995 | 1.230 | (1.23060) | | 0.761 | (0.76533) | |
| 0.999 | 1.449 | | | 1.075 | | |
| $\beta/a = 10$ | | | | | | |
| 0.2 | 0.99816 | | 0.99816 | 0.00030 | | 0.00030 |
| 0.4 | 0.98558 | | 0.98563 | 0.00496 | | 0.00483 |
| 0.6 | 0.95065 | | 0.95099 | 0.02723 | | 0.02484 |
| 0.8 | 0.87024 | | 0.87445 | 0.10440 | | 0.08088 |
| 0.9 | 0.78727 | | 0.80822 | 0.20427 | | 0.13233 |
| 0.95 | 0.7117 | | 0.76452 | 0.3033 | | 0.16632 |
| 0.99 | 0.5737 | | 0.72318 | 0.4851 | | 0.19826 |
| 0.995 | 0.536 | | | 0.538 | | |
| 0.999 | 0.501 | | | 0.596 | | |

of points required to be taken. When $a/b = 0.999$, its boundary points need to reach $M = 36$ and $N = 36$, or more, for convergence.

Through the use of spherical bipolar coordinates, Keh and Chen (1988) obtained semianalytical-seminumerical solutions for the normalized translational and rotational velocities of a dielectric sphere surrounded by an infinitesimally thin electric double layer undergoing electrophoresis parallel to a nonconducting plane wall. These solutions, which can apply to the case of diffusiophoresis of a sphere with relaxation parameter is equal to zero parallel to an impermeable plane wall, are also presented in Table 2.1 for comparison. It can be seen that our collocation solutions for the particle velocities agree excellently with the bipolar-coordinate solutions.

In Appendix C1, I have demonstrated the reflection method to obtain the spherical particles that are parallel to a flat plate, and the analytical solution of diffusion mobility. The motion and the velocity of rotation of particles found is represented in Formula [C1-11a, b], and the results of this calculation are also listed in the table to be compared with the correct numerical results obtained with border access. The results show that the numerical results are very consistent when $\lambda = a/b \leq 0.8$, in addition, when valuing the reflection method with the boundary collocation method regularization mobile velocity, it can be shown that its error is less than 1.1 %. However, the accuracy of Formula [C1-11a, b] decreases with the increase of $\lambda$.

**Table. 2.2** Diffusiophoretic motion and rotating velocity of *spherical particles* parallel to a single plate appear in linear distribution

| $a/b$ | $U/U_0$ | | $-a\Omega/U_0$ | |
|---|---|---|---|---|
| | Exact solution | Asymptotic solution | Exact solution | Asymptotic solution |
| $\beta/a = 0$ | | | | |
| 0.2 | 0.99853 | 0.99853 | 0.00030 | 0.00030 |
| 0.4 | 0.98840 | 0.98846 | 0.00498 | 0.00480 |
| 0.6 | 0.95922 | 0.95993 | 0.02767 | 0.02422 |
| 0.8 | 0.88358 | 0.89274 | 0.10880 | 0.07619 |
| 0.9 | 0.79405 | 0.83125 | 0.21593 | 0.12162 |
| 0.95 | 0.7074 | 0.78952 | 0.3200 | 0.15067 |
| 0.99 | 0.5509 | 0.74938 | 0.4940 | 0.17738 |
| 0.995 | 0.510 | | 0.539 | |
| 0.999 | 0.468 | | 0.584 | |
| $\beta/a = 10$ | | | | |
| 0.2 | 0.99990 | 0.99990 | 0.00030 | 0.00030 |
| 0.4 | 0.99973 | 0.99977 | 0.00494 | 0.00536 |
| 0.6 | 1.00126 | 1.00132 | 0.02715 | 0.03393 |
| 0.8 | 1.00835 | 1.00571 | 0.10687 | 0.14890 |
| 0.9 | 1.01621 | 1.00762 | 0.22178 | 0.28746 |
| 0.95 | 1.0216 | 1.00774 | 0.3556 | 0.39282 |
| 0.99 | 1.0251 | 1.00708 | 0.6874 | 0.50057 |
| 0.995 | 1.036 | | 0.826 | |
| 0.999 | 1.098 | | 1.040 | |

The exact numerical solutions for the normalized velocities $U/U_0$ and $a\Omega/U_0$ of a spherical particle undergoing diffusiophoresis parallel to a plane wall as functions of $a/b$ are depicted in Fig. 2 for various of $\beta/a$. It can be seen that the wall-corrected normalized diffusiophoretic mobility $U/U_0$ of the particle decreases with an increase in $\beta/a$ for the case of an impermeable wall (the boundary condition (2.5) is used), but increases with an increase in $\beta/a$ for the case of a plane wall prescribed with the far-field solute concentration distribution (the boundary condition (2.7) is used), keeping the ratio $a/b$ unchanged. This decrease and increase in the particle mobility becomes more pronounced as $a/b$ increases.

This behavior is expected knowing that the solute concentration gradients on the particle surface near an impermeable wall decrease as the relaxation parameter $\beta/a$ increase and these gradients near a wall with the imposed far-field concentration gradient increase as $\beta/a$ increases (see the analysis in Appendix C1).

When $\beta/a = 1/2$ under the boundary conditions, two different plane walls will have the same particle diffusion mobility. Under this special circumstance, the fluid concentration between the plane walls and particle interactions disappears. As for the flow force effects caused by the presence of the plane walls for particle, the diffusion of the colloidal particles' movable degrees will monotonously decrease with the increase of $a/b$.

**Fig. 2.2  a** Plots of the normalized translational velocity $U = U_0$ of a *spherical particle* undergoing diffusiophoresis parallel to a plane wall versus the separation parameter $a = b$. **b** Figure of velocity $a\Omega/U_0$ to $a/b$ of colloidal particles parallel diffusiophoresis of the single plate

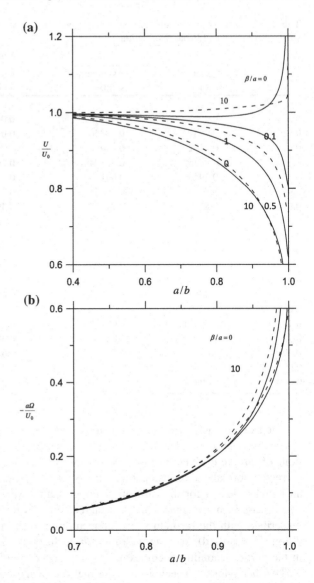

When examining Tables 2.1 and 2.2 and Fig. 2.2a, we find an interesting phenomenon: the plate that is qualitatively dissolved is nonpenetrable, while the polarization coefficient $\beta/a$ is minimum (such as $\beta/a = 0$) of the case, when the $a/b$ decreases, the diffusion of the particles' movable degrees with $a/b$ increases while decreasing extremely to a small value, and later with the increase of the $a/b$, it increases. When the gap between the particles and the plane walls is small enough, the velocity of motion of the particles is even larger than in the situation when plane walls are present. For example, when $\beta/a = 0$ and $a/b = 0.999$, the velocity of motion of the colloidal particles will be 45 % faster than the values in the absence of plane walls. In the case of higher $\beta/a$, the velocity of motion of the plane walls particles is parallel to the fluid which cannot be penetrated; with increase of $a/b$, the velocity decreases monotonically. In the case of the flat surface as a fluid distribution of linear polarization, parameter $\beta/a$ is relatively larger (such as the $\beta/a = 10$); when $a/b$ is smaller, the diffusion of the particles' movable degrees with $a/b$ drops to a small value, and then increases with the increase of $a/b$. When the gap between the particles and the plane walls is small enough, the velocity of motion of the particles will also be greater than the situation when there are no plane walls. In the case when $\beta/a$ is smaller, the velocity of motion of the particles' parallel fluid distribution plate is linear; with the increase of $a/b$, the velocity decreases monotonously.

For the interesting case that $U/U_0$ is not monotone decreasing, it is understood that due to the flow force of resistance effect and fluid concentration gradient growth of the interactions of reasoning result, for cause when plate for fluid does not penetrate and polarization parameter $\beta/a$ is extremely small, and when the plate for fluid is in a linear distribution, while polarization parameter $\beta/a$ larger circumstances, the particle motion velocity is increased with $a/b$, which appears to decrease first and then increase (Please check the sentence for clarity in meaning). Through the reflection method, the $U/U_0$ (formula C1-11 a) is the same as the situation described in the figure.

In the same geometry situation, spherical colloidal particles as diffusion in Tables 2.1 and 2.2 and Fig. 2.2b are influenced by body force field, (such as gravitational field) and the influence caused by the motion of rotation, while the directions of rotation are opposite. This is a comparison with a thin electric double layer of charged particle parallel to an insulating plate of electrophoresis, which is quite approximate (see Keh and Chen 1988). For any polarization parameter $\beta/a$, particle diffusiophoresis is normalized rotational velocity as $a\Omega/U_0$ for $a/b$ of the monotone increasing function, however, when $a/b$ is not big, the influence of $\beta/a$ to $a\Omega/U_0$ is not significant.

O 'Neill (1964) and Ganatos et al. (1980) had respectively used spherical bipolar coordinates and boundary take point method to solve the subcritical flow motion parallel to the infinite plane walls of a single spherical particle which was influenced by body force $Fe_x$. Comparing the spherical particle by gravitational field (at this time, $U_0 = F/6\pi\eta a$) and the effect of diffusion force, we can find the situation of particles in diffusiophoresis when they are influenced by the plane walls which is far less than that of settlement motion.

In parenthesis is the calculated value based on the results of double spherical coordinates according to Keh and Chen (1988).

In Fig. 2.2b, the colloidal particles parallel to the single plate of the diffusion rotational velocity display $a\Omega/U_0$ to $a/b$ mapping (the solid line for solute does not penetrate the plate, while the dashed line for solute is a linear distribution of plate).

**Table 2.3** Diffusiophoretic motion of *spherical particles* parallel to two single plane walls

| $a/b$ | $U/U_0$ | | | |
| --- | --- | --- | --- | --- |
| | $\beta/a = 0$ | | $\beta/a = 10$ | |
| | Exact solution | Asymptotic solution | Exact solution | Asymptotic solution |
| *For impermeable plane walls* | | | | |
| 0.2 | 0.99796 | 0.99796 | 0.99470 | 0.99470 |
| 0.4 | 0.98597 | 0.98617 | 0.96038 | 0.96083 |
| 0.6 | 0.96339 | 0.96661 | 0.87952 | 0.88817 |
| 0.8 | 0.94684 | 0.96326 | 0.74486 | 0.81008 |
| 0.9 | 0.96662 | 0.98325 | 0.64927 | 0.79933 |
| 0.95 | 1.0163 | 1.00270 | 0.5839 | 0.81018 |
| 0.99 | 1.2243 | 1.02412 | 0.4935 | 0.82992 |
| 0.995 | 1.363 | | 0.481 | |
| 0.999 | 1.718 | | 0.494 | |
| *For plane walls prescribed with the far-field concentration profile* | | | | |
| 0.2 | 0.99586 | 0.99587 | 0.99832 | 0.99832 |
| 0.4 | 0.96931 | 0.96975 | 0.98877 | 0.98899 |
| 0.6 | 0.90602 | 0.91448 | 0.97215 | 0.97605 |
| 0.8 | 0.79069 | 0.85485 | 0.96200 | 0.98514 |
| 0.9 | 0.69390 | 0.84471 | 0.97401 | 1.01389 |
| 0.95 | 0.6184 | 0.85078 | 0.9974 | 1.03838 |
| 0.99 | 0.5075 | 0.86316 | 1.0550 | 1.06415 |
| 0.995 | 0.490 | | 1.113 | |
| 0.999 | 0.497 | | 1.256 | |

## 2.2.2  Diffusiophoresis of Particle Parallel to Two Plane Walls

Table 2.3 compares when a spherical particle is placed in two parallel plane walls (when $c = b$), and is parallel to two plane walls, the diffusiophoresis for different polarization ratio $\beta/a$ and separating parameters (separation Parameter) $a/b$ in two different boundary conditions of the plane walls, the outcomes of which are under the boundary condition using point method, and with the reflection method for approximate results proved by each other (see Appendix C type (C1-20)).

Approximate to the motion situation of single particle parallel to the plate, in order to take the boundary collocation method for correct outcomes and the reflection method for approximate results (Formula (C1-20)), when $\lambda \leq 0.6$, the two kinds of calculations are very approximate. But when $\lambda \geq 0.8$, with reflection method the approximate result is considerably different. Generally speaking, Formula (C1-20) overestimates the particle diffusion velocity. Comparing Table 2.3 with Tables 2.1 and 2.2, we can see that when $a/b$ is relatively small and as a single plate of the boundary effect directly added into, it will underestimate the two-plate boundary effect; however, when $a/b$ is very large and as a single

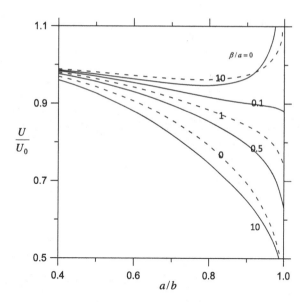

**Fig. 2.3** Plots of the normalized diffusiophoretic mobility $U = U_0$ of a *spherical particle* migrating on the median plane between two parallel plane walls (with $c = b$) versus the separation parameter $a = b$ for several values of $r = a$

plate boundary effect of direct additive into, it will overestimate the two-plate boundary effect.

For different polarization ratios $\beta/a$, with separate parameters $a/b$ of the particle regularization diffusion, the shift motion velocity $U/U_0$ and rotational velocity $a\Omega/U_0$ to boundary take point method for numerical results are shown in Fig. 2.3. According to the figure, it is known that the plate in which the fluid cannot penetrate, its regularization diffusiophoresis velocity $U/U_0$ will also increase with the polarization ratio $\beta/a$ gradually reducing. However, in the plate for linear concentration distribution of cases, the normalized diffusiophoresis velocity $U/U_0$ will also increase when the polarization ratio $r/a$ increasing. Approximately, when plate for fluid does not penetrate, and polarization ratio $\beta/a$ is minimum (such as $\beta/a = 0$), where $a/b$ is relatively small, the particle of the diffusion mobility increases with the increase of $a/b$ and declines to a minimum, then when $a/b$ increases, it increases. In addition, when the gap between particles and plane walls is small enough, the particle's motion velocity is also larger than the situation without plane walls. Therefore, when the gap between particles and plane walls is extremely small, fluid concentration gradient effect of growth will be larger than the power of flow resistance effect, therefore causing the particle motion velocity to increase. Through the reflection method for approximate results (type (C1-20)), its trend corresponds with the outcomes obtained from the boundary take point method.

Comparing Figs. 2.3 and 2.2a, we can conclude that when the second plate is added, it does not necessarily enhance the particle due to influence of diffusion swimming velocity (even if two plane walls and particles are of equal distance). Because when joining the second plate, the flow force of resistance effect and the fluid concentration gradient effect of growth although increased have different

**Fig. 2.4  a** Plots of the normalized translational velocities $U = U_0$ of a *spherical particle* undergoing diffusiophoresis parallel to two plane walls versus the ratio $b = (b + c)$. **b** When $\beta/a = 1/2$, the diffusiophoresis rotational velocity of colloidal particles $a\Omega/U_0$, make figure for $b/(b + c)$ between two parallel plane walls

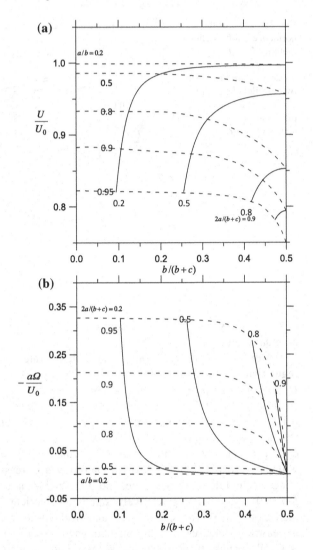

degrees, so that the total influence does not necessarily enhance the particles of the diffusiophoresis velocity influence, which is the so-called "Phoretic Migration Enhancing Effect (PMEE)."

Colloidal particles are located at any position between two plane walls, for different separation parameters $a/b$, when the polarization parameter $\beta/a = 1/2$ (now for two kinds of different boundary conditions of plane walls, they have the same numerical results), particle regularization diffusiophoresis velocity $U/U_0$ and rotational velocity of $a\Omega/U_0$ numerical results as shown in Fig. 2.4.

The dotted line in the figure represents distance between the fixed one of the plane walls and the particles ($a/b = $ constant), and the influence caused by the change of another plate (in $z = c$) for the diffusiophoresis of colloidal particles. The line represents the fixed distance between two plane walls ($2a/(b + c)$ = constant); when the particle is located at different locations between two plane walls, the influence they cause is to colloid particle diffusiophoresis. From Fig. 2.4a, we know that under this circumstance, plate net effect will reduce particle diffusion velocity $U/U_0$. When $2a/(b + c)$ is a fixed value, and particle is located in the middle between two plane walls ($c = b$), the drag force is minimum, and it has the largest mobile velocity (the rotational velocity is zero). While when the particle gradually moves toward a plate (when $b/(b + c)$ decrease), the drag force of fluid increases, and the motion velocity decreases but the rotational velocity increases.

With a fixed particle and at a distance from one of the plane walls (when $a/b$ is a fixed value), the existence of another plate will reduce particle motion velocity and rotating velocity, and with the particles and another flat gradually moving toward each other (when $b/(b + c)$ gradually increases), particle motion velocity and rotating velocity gradually decreases as well.

On the other hand, for some cases such as the diffusiophoresis of a colloidal sphere with a small value of $\beta/a$ parallel to two impermeable plane walls or with a large value of $\beta/a$ parallel to two plates prescribed with the far-field solute concentration distribution, the net wall effect can increase the diffusiophoretic mobility of the particle relative to its isolated value.

If we compare different $2a/(b + c)$ cases with each other, it can be found that when the particle is located in the middle of two plane walls, at this time there is relatively maximum particle velocity value; while when the particle is near any one of the plane walls the relative velocity decreases. What is worth mentioning is that in fixed particle and the distance of a flat (fixed $a/b$ value), another plate to the effect of particle motion velocity is not monotone function, however, for the sake of conciseness, I do not provide a figure to illustrate this.

Ganatos et al. (1980) used the boundary collocation method to solve when a single spherical particle is located at any position in between two flats, parallel plane walls will display falling motion. Compared with this chapter, we can see that in general cases, the effect of plane walls diffusiophoresis is much smaller than a falling motion.

## 2.3 Conclusions

This study considers single spherical colloidal particles in the case of low Reynolds number and low picogram number of columns, parallel in a single infinite plate or infinite plate of diffusiophoretic motion behavior, respectively, taking point method (boundary collocation method) and reflection method for solving particle swimming velocity and comparing particle phoretic motion in the case.

In this chapter, I consider a single spherical colloidal particle diffusion in the electrolyte solution. The boundary conditions of the plates are solute impermeable and solute linear distribution of the two cases. The results found that the diffu-siophoresis boundary effects varied. Generally speaking, the diffusiophoretic velocity of separation parameters has a monotonic decreasing function. However, when a/b is close to 1, under different parameters of polarization ($= \beta/a$), the boundary effects of plates could accelerate or decelerate motion velocity of par-ticles (compared with a single particle at infinity condition); this phenomenon of the acceleration effect of concentration gradient and the effect of viscosity of fluid mechanics should make a situation of competition for both.

# References

Anderson, J.L., Prieve, D.C.: Diffusiophoresis caused by gradients of strongly adsorbing solutes. Langmuir **7**, 403 (1991)

Ganatos, P., Weinbaum, S., Pfeffer, R.: A strong interaction theory for the creeping motion of a sphere between plane parallel boundaries. Part 2. Parallel motion. J. Fluid Mech. **99**, 755 (1980)

Keh, H.J., Chen, S.B.: Electrophoresis of a colloidal sphere parallel to a dielectricplane. J. Fluid Mech. **194**, 377 (1988)

Keh, H.J., Jan, J.S.: Boundary effects on diffusiophoresis and electrophoresis: Motion of a colloidal sphere normal to a plane wall. J. Colloid Interface Sci. **183**, 458 (1996)

O'Brien, R.W.: The solution of the electrokinetic equations for colloidal particles with thin double layers. J. Colloid Interface Sci. **92**, 204 (1983)

O'Neill, M.E.: A slow motion of viscous liquid caused by a slowly moving solid sphere. Mathematika **11**, 67 (1964)

O'Brien, V.: Form factors for deformed spheroids in Stokes flow. AIChE J. **14**, 870 (1968)

# Chapter 3
# Osmophoretic Motion of the Spherical Vesicle Particle Parallel to Plane Walls

**Abstract** This chapter concerns the parallel penetration motion of the single spherical vesicle particles driven by the settled concentration gradient. The boundary conditions of the plane wall have two situations: the solute that cannot be penetrated; and the linear distribution of the solvent quality. As for the boundary effects of the plate for osmophoretic bathing, one is from the interaction generated between the particles and the plate, and another by the viscous effects of the fluid. This chapter makes boundary collocation method to calculate the vesicle particle penetration in various situations of motion velocity and reflection, compares the reflection method, and makes sure their results are consistent. The border effect of plate penetration motion is determined by the characteristics of the particles, as well as the relative distance from the plate and also the boundary conditions of the solute on the plate.

## 3.1 Theoretical Analysis

The chapter considers the radius of spherical vesicle particles $a$, osmophoretic motion which between two parallel plane walls is affected by external solute concentration gradient. The distance from the center of the spherical vesicle particle to two plates is $b$ and $c$ respectively, as shown in Fig. 3.1, where $(x, y, z)$ $(\rho, \varphi, z)$ and $(r, \theta, \varphi)$ represent the Cartesian coordinates, cylindrical coordinates, and spherical coordinate systems, respectively, while the center of particle is taken as origin. In infinite situation, solute concentration $C_\infty(x)$ performs a linear distribution, and the concentration gradient is $E_\infty e_x$ ($=\nabla C_\infty$, where $E_\infty$ is positive) while $e_x$, $e_y$ and $e_z$, are the three unit vectors of the Cartesian coordinates. We assume the spherical particles remain in shape, and the solute internal and external are unable to penetrate the vesicle films. Hence, it is theoretically estimating correction of osmophoretic motion velocity of single particle in formula (1.3a, b) in the presence of plates. Therefore, the distribution of solute concentration must be available prior to solving the velocity field of fluid.

P.-Y. Chen, *The Application of Biofluid Mechanics*, SpringerBriefs in Physics, DOI: 10.1007/978-3-642-44952-9_3, © The Author(s) 2014

**Fig. 3.1** Coordinate graph of
osmophoresis of single
spherical vesicle particle in
parallel plane walls

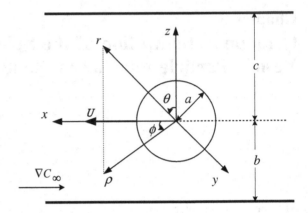

## 3.1.1 Distribution of Solute Concentration

While Reynolds number and Peclet number are so small that they can be negligible, the osmophoretic motion status can be regarded as a quasi steady-state system. The conservation equations compliant with internal and external distribution of solute concentration of spherical vesicle are, respectively, as follows:

$$\nabla^2 C = 0 \quad (r \geq a) \tag{3.1a}$$

$$\nabla^2 C_1 = 0 \quad (r \leq a) \tag{3.1b}$$

where $C$ and $C_1$ represent the external and internal distributions of solute concentration of spherical vesicle, respectively. Because the radius of spherical vesicle is much larger than its thickness, equation $r = a$ can represent surfaces of inner and outer layers of it. Hence (Anderson 1983; Keh and Yang 1993a):

$$r = a: \quad \frac{\partial C_1}{\partial r} = \frac{\bar{\kappa}}{a}[C - C_0 - (C_1 - \bar{C})] \tag{3.2a}$$

$$\frac{\partial C}{\partial r} = \frac{\kappa}{a}[C - C_0 - (C_1 - \bar{C})] \tag{3.2b}$$

In the above formula, the definitions of $k$ and $\bar{k}$ are shown in formula (1.3c, d).

The far-filed solute concentration exists without disturbance, and the fluid hereby also stands still. Hence,

$$z = c, \ -b: \quad \frac{\partial C}{\partial z} = 0 \tag{3.3}$$

$$\rho \to \infty: \quad C \to C_\infty = C_0 - E_\infty x \tag{3.4}$$

where the formula (3.3) is boundary conditions of two plane walls of impermeable solute. For boundary conditions of two plane walls of linear distributed solute concentration, the formula (3.3) should be revised as

$$z = c, -b: \quad C = C_\infty \tag{3.5}$$

Since the governing equation and boundary conditions are all linear, the distribution of solute concentration $C$ could be performed as

$$C = C_w + C_p \tag{3.6}$$

where the formula (3.1a) due to disturbance of presence of plates, $C_w$, can find its general solution by Double Fourier integral with its Cartesian coordinates, which is coupled with the far-filed concentration distribution without disturbance and could be performed as

$$C_w = C_0 - E_\infty x - E_\infty \int_0^\infty \int_0^\infty (Xe^{\kappa z} + Ye^{-\kappa z}) \sin(\hat{\alpha} x) \cos(\hat{\beta} y) d\hat{\alpha} \, d\hat{\beta} \tag{3.7}$$

where $X$ and $Y$ are functions to be determined, and $k = (\hat{\alpha}^2 + \hat{\beta}^2)^{1/2}$. $C_p$ satisfies the formula (3.1a, b), the general solution spherical coordinate caused by disturbances of existing spherical vesicles a spherical harmonic function:

$$C_p = -E_\infty \sum_{n=1}^\infty R_n r^{-n-1} P_n^1(\mu) \cos \phi \tag{3.8}$$

where $P_n^1$ is associated Legendre function, while $\mu$ represents $\cos \theta$ in order to be concise, and $R_n$ is an unknown coefficient compared with the unknown coefficient. The concentration distribution, $C$, shown in formula (3.6–3.8) already satisfies the boundary conditions of infinite places in formula (3.4). The general solution of concentration field of spherical vesicle is shown as below:

$$C_1 = \bar{C} - E_\infty x - E_\infty \sum_{n=1}^\infty \bar{R}_n r^n P_n^1(\mu) \cos \phi \tag{3.9}$$

The distribution of solute concentration, $C$, of formula (3.6–3.8) is substituted into the boundary condition of formula (3.3) (or 3.5); after Fourier sine and cosine transforms of $X$ and $Y$, they can be converted into the equation represented by $R_n$. Its general solution can be substituted into formula (3.6) by then, the distribution of solute concentration, $C$, can be performed as integral state of the modified Bessel functions of the second kind as below:

$$C = C_0 - E_\infty x - E_\infty \sum_{n=1}^\infty R_n \delta_n^{(1)}(r, \mu) \cos \phi \tag{3.10}$$

For the detailed definition of function $\delta_n^{(1)}(r, \mu)$, please refer to formula (D1) in Appendix D. By substituting formula 3.9 and 3.10 into the boundary conditions formula (3.2a, b), it can be obtained as follows:

$$\frac{\bar{\kappa}}{a}\sum_{n=1}^{\infty} R_n\delta_n^{(1)}(a, \mu) - \sum_{n=1}^{\infty}\bar{R}_n(\bar{\kappa}+n)a^{n-1}P_n^1(\mu) = (1-\mu^2)^{1/2} \qquad (3.11a)$$

$$\sum_{n=1}^{\infty} R_n\left[\frac{\kappa}{a}\delta_n^{(1)}(a, \mu) - \delta_n^{(2)}(a, \mu)\right] - \kappa\sum_{n=1}^{\infty}\bar{R}_n a^{n-1}P_n^1(\mu) = (1-\mu^2)^{1/2} \qquad (3.11b)$$

For the details of function $\delta_n^{(1)}(r, \mu)$ please refer to formula (D2) in Appendix D. The integrals of $\delta_n^{(1)}$ and $\delta_n^{(2)}$ are obtained via numerical integration.

It needs infinite numbers of undetermined coefficients, $R_n$, to really satisfy the boundary conditional formula (3.11a, b). However, the method of boundary collocation method allows us to abridge infinite series formula (3.11a, b) into a limited series and to take limited numbers of points on the surface of each particle to satisfy the boundary conditions (O'Brien 1968; Ganatos et al. 1980; Keh and Jan 1996). For each infinite series formula $\sum_{n=0}^{\infty}$, after obtaining its first $M$ item, there should be $M$ of unknown coefficients $R_n$ needed to be obtained. Using the satisfactory $M$ of different values of $\theta_i$ on the surface of each spherical particle within formula (3.11a, b) to generate $M$ of formulas, which can be utilized to obtain undetermined coefficients, $R_n$.

## 3.1.2 Distribution of Fluid Velocity

From the distribution of the solute concentration obtained in the previous section, this system can be used to further calculate the distribution of fluid velocity. Fluid is assumed as incompressible Newtonian fluid, based on the moving of the semipermeable membrane in creeping flow; the flow field by the Stokes equation is performed as below:

$$\eta\nabla^2 v - \nabla p = 0 \qquad (3.12a)$$

$$\nabla \cdot v = 0 \qquad (3.12b)$$

where $v$ represents velocity distribution of fluid, and $p$ represents its distribution of stress.

The boundary conditions of fluid velocity on particle surfaces and in infinite places are (Anderson 1983; Keh and Yang 1993b):

$$r = a : \quad v = U + a\boldsymbol{\Omega} \times e_r + L_pRT[C - C_0 - (C_1 - \bar{C})]e_r \qquad (3.13)$$

$$z = c, -b : \quad v = 0 \qquad (3.14)$$

$$\rho \to \infty : \quad v = 0 \qquad (3.15)$$

where $e_r$, $e_\theta$ and $e_\varphi$ are unit vectors of spherical coordinates, as for $U = Ue_x$ and $\Omega = \Omega e_y$ are respectively moving velocity and rotation velocity of osmophoretic motion of spherical vesicle. The moving velocity of spherical vesicle is still parallel to solute concentration gradient in asymmetrical situation (while $b \neq c$) due to ignoring effects of inertial term.

The general solution of velocity satisfying Stroke formula (formula 3.12a, b) and boundary conditions (3.14) and (3.15) could be performed as follows:

$$v = v_x e_x + v_y e_y + v_z e_z \tag{3.16}$$

As for

$$v_x = \sum_{n=1}^{\infty} [A_n(A_n' + \alpha_n') + B_n(B_n' + \beta_n') + C_n(C_n' + \gamma_n')] \tag{3.17a}$$

$$v_y = \sum_{n=1}^{\infty} [A_n(A_n'' + \alpha_n'') + B_n(B_n'' + \beta_n'') + C_n(C_n'' + \gamma_n'')] \tag{3.17b}$$

$$v_z = \sum_{n=1}^{\infty} [A_n(A_n''' + \alpha_n''') + B_n(B_n''' + \beta_n''') + C_n(C_n''' + \gamma_n''')] \tag{3.17c}$$

where the labels $A_n$, $B_n$, $C_n$, $\alpha_n$, $\beta_n$ and $\gamma_n$ all contain associated Legendre function using $\mu$ (or $\cos \theta$) as even number, and for its detailed definition please refer to formula (2.6) and the formula (C1) in Ganatos et al. (1980) (see the Appendix D of this book), and $A_n$, $B_n$ and $C_n$ are undetermined coefficients.

In order to satisfy its boundary conditions, formulas (3.10) and (3.16) are substituted into (3.13) to obtain formulas as follows:

$$\sum_{n=1}^{\infty} [A_n(A_n' + \alpha_n') + B_n(B_n' + \beta_n') + C_n(C_n' + \gamma_n')]$$

$$= U + a\Omega\mu - L_p RTE_\infty \sum_{n=1}^{\infty} [R_n \delta_n^{(1)}(a, \mu) - \bar{R}_n a^n P_n^1(\mu)](1 - \mu^2)^{1/2} \cos^2 \phi \tag{3.18a}$$

$$\sum_{n=1}^{\infty} [A_n(A_n'' + \alpha_n'') + B_n(B_n'' + \beta_n'') + C_n(C_n'' + \gamma_n'')]$$

$$= -L_p RTE_\infty \sum_{n=1}^{\infty} [R_n \delta_n^{(1)}(a, \mu) - \bar{R}_n a^n P_n^1(\mu)](1 - \mu^2)^{1/2} \sin \phi \cos \phi \tag{3.18b}$$

$$\sum_{n=1}^{\infty} [A_n(A_n''' + \alpha_n''') + B_n(B_n''' + \beta_n''') + C_n(C_n''' + \gamma_n''')]$$

$$= -a\Omega(1 - \mu^2)^{1/2} \cos \phi - L_p RTE_\infty \sum_{n=1}^{\infty} [R_n \delta_n^{(1)}(a, \mu) - \bar{R}_n a^n P_n^1(\mu)] \mu \cos \phi$$

$$(3.18c)$$

After detailed observation on formula (3.18a–c), we can discover that while using the boundary collocation method at the boundary of the ball surface, $r = a$, all the associated verticals are not related to the selection of the value of $\varphi$. Therefore, the formula (3.18a–c) is compliant with $N$ of different values of $\theta i$ on each surface of spherical particle (values of $\theta$ is between 0 and $\pi$) resulting in $3N$ of linear equation, and can be applied to solve $3N$ of unknowns, $A_n$, $B_n$ and $C_n$, and the distribution of fluid field could be obtained smoothly when $N$ is large enough.

### 3.1.3  The Derivation of Osmophoresis of Particles

The drag force and torque of fluid upon spherical vesicle particles could be performed as (Ganatos et al. 1980):

$$F = -8\pi\eta A_1 \ e_x \qquad (3.19a)$$

$$T = -8\pi\eta C_1 \ e_y \qquad (3.19b)$$

From the above formula it could be figured that there are only contributions of lower terms as $A_1$ and $C_1$ for drag force and torque upon spherical particles in formula (3.17a–c).

Because the particles are freely floating in solution, the particles have received zero of net force and net torque. Applying this limitation into formula (3.19a, b), the result is obtained as

$$A_1 = C_1 = 0 \qquad (3.20)$$

By combining the $3N$ of linear equations generated in formula (3.18a–c) with the associated vertical (3.20), we can successfully obtain the moving velocity and rotation velocity of particles.

## 3.2  Results and Discussion

This section discusses the application of boundary collocation method in solving calculation results of osmophoretic motion of single spherical vesicle particle that is parallel to the two plane walls.

In Tables 3.1 and 3.2, respectively, the resulted numeric values of different parameters, $k, \bar{k}$ and $a/b$, are compared using the boundary collocation methods, under two different plates boundary conditions, through the osmophoretic motion of spherical vesicle particle that is parallel to a single plate (when $c \to \infty$), and processing mutual verification with approximate results generated from reflection method.

In Tables 3.1 and 3.2, both using the numerical calculation of the boundary collocation method converges to present effective digits. Convergence rate is related to the values of $a/b$. The larger the values are, the slower the convergence rate is, and also the more points required to be accessed. When $a/b = 0.999$, the boundary collocation method must reach $M = 36$ and $N = 36$, or above, to be able to start convergence.

In Appendix C2, I illustrate the spherical vesicle particle with reflection method and obtain approximate analytical solution of osmophoresis in detail, and the velocity of rotation of particles found in formula (C2-11a, b). The results of this calculation are also listed in the table; the boundary collocation method to obtain the correct value results are comparable to each other. The results show that when $\lambda \leq 0.8$ boundary collocation method obtained from the results agree well with the value of motion velocity value obtained by the reflection method, and the error is less than 3.2 %. However, when the value of $\lambda$ is greater, the accuracy of formula (C2-11a, b) will be lower.

For different parameters $\kappa$ and $\bar{\kappa}$, the moving velocity with which the particle regularization diffusiophoresis $U/U_0$ with rotational velocity $a\Omega/U_0$, the numerical results are shown in Figs. 3.2 and 3.3. The figures show that, at the parallel plane walls (boundary conditions of the formula 3.5), the regularization diffusiophoresis velocity in the case of solute impermeable, the parameter $U/U_0$ will reduce gradually with the reducing values of parameters $\kappa$ and $\bar{\kappa}$. However, on the plates of linear case of the concentration distribution (boundary conditions of the formula 3.7), the regularization osmophoretic motion velocity $U/U_0$ will increase with the reducing values of parameters $\kappa$ and $\bar{\kappa}$.

When the separation parameter $a/b$ increases, I hereby provide some further demonstrations of the changing mobility of vesicle particles: when the plate is impermeable to the solute (boundary condition formula 3.5), around the particles of the solute concentration gradient with increased) $\kappa/(1 + \bar{\kappa})$ decline in plate. The cases of linear distribution of solute concentration, around the particles of the solute concentration gradient increase as when values of $\kappa/(1 + \bar{\kappa})$ rises (see Appendix C2 for further analysis).

When $\kappa = 1 + \bar{\kappa}$, with different boundary conditions, the two different plates will have the same effects on diffusiophoresis of particles. These special conditions, the cross interaction of solute between the plate and vesicle particles, disappear. Since the only presence of plates leading the growth of flow force effect to vesicle particles, the mobility of osmophoresis of vesicle particles will increase monotonically with increasing values of $a/b$.

**Table 3.1** Normalized translational and rotational velocities of a spherical vesicle undergoing osmophoresis parallel to a single impermeable plane wall

| $a/b$ | $U/U_0$ | | $a\Omega/U_0$ | |
|---|---|---|---|---|
| | Exact solution | Asymptotic solution | Exact solution | Asymptotic solution |
| $\kappa = \bar{\kappa} = 0$ | | | | |
| 0.2 | 1.00245 | 1.00244 | 0.00060 | 0.00061 |
| 0.4 | 1.01905 | 1.01893 | 0.00998 | 0.01052 |
| 0.6 | 1.06651 | 1.06501 | 0.05604 | 0.06424 |
| 0.8 | 1.19518 | 1.17331 | 0.23161 | 0.27075 |
| 0.9 | 1.37549 | 1.27325 | 0.50444 | 0.51321 |
| 0.95 | 1.6066 | 1.34153 | 0.8468 | 0.69553 |
| 0.99 | 2.3358 | 1.40749 | 1.8148 | 0.88087 |
| 0.995 | 2.647 | | 2.245 | |
| 0.999 | 3.049 | | 2.838 | |
| $\kappa = 10\ \bar{\kappa} = 0$ | | | | |
| 0.2 | 1.00118 | 1.00118 | 0.00060 | 0.00059 |
| 0.4 | 1.00839 | 1.00830 | 0.00976 | 0.00884 |
| 0.6 | 1.02411 | 1.02406 | 0.05141 | 0.03568 |
| 0.8 | 1.04507 | 1.05284 | 0.17650 | 0.05681 |
| 0.9 | 1.04477 | 1.07731 | 0.30197 | 0.02530 |
| 0.95 | 1.0203 | 1.09409 | 0.3877 | −0.01687 |
| 0.99 | 0.9038 | 1.11055 | 0.4433 | −0.06996 |
| 0.995 | 0.853 | | 0.436 | |
| 0.999 | 0.783 | | 0.431 | |
| $\kappa = 0\ \bar{\kappa} = 10$ | | | | |
| 0.2 | 1.00245 | 1.00245 | 0.00060 | 0.00061 |
| 0.4 | 1.01921 | 1.01906 | 0.01002 | 0.01094 |
| 0.6 | 1.06883 | 1.06653 | 0.05698 | 0.07153 |
| 0.8 | 1.21930 | 1.18185 | 0.24760 | 0.32536 |
| 0.9 | 1.47298 | 1.29055 | 0.58685 | 0.63777 |
| 0.95 | 1.8744 | 1.36546 | 1.1032 | 0.87739 |
| 0.99 | 3.7155 | 1.43814 | 3.2605 | 1.12360 |
| 0.995 | 4.804 | | 4.663 | |
| 0.999 | 6.458 | | 7.126 | |

Observing Tables 3.1, 3.2, Figs. 3.2a and 3.3a in detail, we find an interesting phenomenon: when the plate is solute impermeable, in the case of $\kappa \gg 1 + \bar{\kappa}$, and the value of $a/b$ is smaller, the mobility of osmophoresis of particles will increase to a maximum value with increasing values of $a/b$, and will decrease with constantly increasing values of $a/b$ by then. If the particle-plate gap is small enough, the velocity of particle motion will be smaller than it is with the absence of plates. For instance, while $k = 10$, $\bar{k} = 0$ and $a/b = 0.999$; the velocity of vesicle particle motions will be 22 % slower than that of absence of plates. In the case of $k \leq 1 + \bar{k}$, velocity of particles parallel to solute impermeable plates will increase monotonically with increasing values of $a/b$. When the plate is the solute linear distribution,

**Table 3.2** Normalized translational and rotational velocities of a spherical vesicle undergoing osmophoresis parallel to a single plane wall prescribed with the far-field solute concentration profile

| $a/b$ | $U/U_0$ | | $a\Omega/U_0$ | |
|---|---|---|---|---|
| | Exact solution | Asymptotic solution | Exact solution | Asymptotic solution |
| $\kappa = \bar{\kappa} = 0$ | | | | |
| 0.2 | 1.00144 | 1.00144 | 0.00060 | 0.00060 |
| 0.4 | 1.01059 | 1.01048 | 0.00982 | 0.00941 |
| 0.6 | 1.03357 | 1.03291 | 0.05280 | 0.04543 |
| 0.8 | 1.08412 | 1.08064 | 0.19413 | 0.12984 |
| 0.9 | 1.14050 | 1.12400 | 0.36882 | 0.19186 |
| 0.95 | 1.1987 | 1.15396 | 0.5393 | 0.22633 |
| 0.99 | 1.3237 | 1.18323 | 0.8565 | 0.25464 |
| 0.995 | 1.357 | | 0.952 | |
| 0.999 | 1.382 | | 1.068 | |
| $\kappa = 10\ \bar{\kappa} = 0$ | | | | |
| 0.2 | 1.00270 | 1.00270 | 0.00060 | 0.00061 |
| 0.4 | 1.02130 | 1.02115 | 0.01004 | 0.01109 |
| 0.6 | 1.07668 | 1.07431 | 0.05753 | 0.07399 |
| 0.8 | 1.24246 | 1.20367 | 0.25303 | 0.34378 |
| 0.9 | 1.50675 | 1.32514 | 0.59729 | 0.67978 |
| 0.95 | 1.8786 | 1.40858 | 1.0821 | 0.93873 |
| 0.99 | 3.0347 | 1.48937 | 2.5159 | 1.20547 |
| 0.995 | 3.479 | | 3.118 | |
| 0.999 | 3.973 | | 3.899 | |
| $\kappa = 0\ \bar{\kappa} = 10$ | | | | |
| 0.2 | 1.00144 | 1.00143 | 0.00060 | 0.00060 |
| 0.4 | 1.01044 | 1.01035 | 0.00979 | 0.00899 |
| 0.6 | 1.03136 | 1.03139 | 0.05191 | 0.03814 |
| 0.8 | 1.06415 | 1.07211 | 0.18097 | 0.07523 |
| 0.9 | 1.07677 | 1.10670 | 0.31581 | 0.06730 |
| 0.95 | 1.0690 | 1.13003 | 0.4184 | 0.04447 |
| 0.99 | 1.0131 | 1.15258 | 0.5419 | 0.01191 |
| 0.995 | 0.987 | | 0.567 | |
| 0.999 | 0.947 | | 0.605 | |

while in the case of $\kappa \ll 1 + \bar{\kappa}$, and the value of $a/b$ is smaller, the mobility of osmophoresis of particles will increase to a maximum value with increasing values of $a/b$, and will decrease with constantly increasing values of $a/b$ by then. In the case of $k \geq 1 + \bar{k}$, velocity of osmophoresis of particles parallel to solute impermeable plates will increase monotonically with increasing values of $a/b$.

For the interesting case of the $U/U_0$ is not monotonic function of $a/b$, it is understandable that the results of interactions between force growth generated by plates and solute concentration is the major reason which caused the prior increasing and decreasing of the velocity of particles motion with increasing

**Fig. 3.2  a** Figure of
osmophoretic motion velocity
of $a\Omega/U_0$ to $a/b$ of a single
plate parallel to a plate when
$\kappa = 10$, **b** Figure of rotational
motion velocity of $a\Omega/U_0$ to
$a/b$ of the vesicle particles
parallel to a single plate when
$\bar{\kappa} = 10$

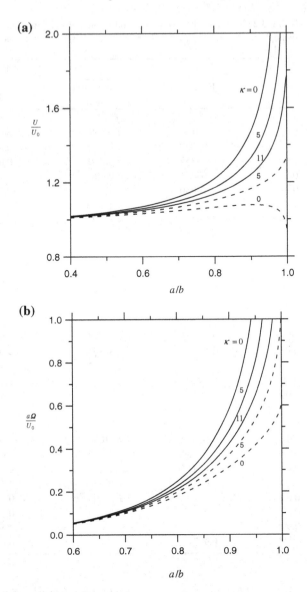

values of $a/b$, while $k/(1 + \bar{k})$ is larger in plates with solute impermeable, and
while $k/(1 + \bar{k})$ is smaller in plates with linear distribution of solute concentration.
The $U/U_0$ obtained by reflection method (formula C2-11a) is compliant with the
situation prescribed in Figs. 3.2a and 3.3a.

In Tables 3.1, 3.2, Figs. 3.2b and 3.3b, the rotation of the spherical vesicle
particles is generated by their diffusiophoresis. Under the same geometric conditions, affection of body-force fields (such as gravity) and the rotation direction of

**Fig. 3.3** **a** Figure of osmophoresis velocity of $U/U_0$ to $a/b$ of the vesicle particles parallel to a single plate when $\kappa = 10$, **b** Figure of rotational motion velocity of $a\Omega/U_0$ to $b/(b+c)$ of the vesicle particles parallel to a single plate when $\kappa = 10$

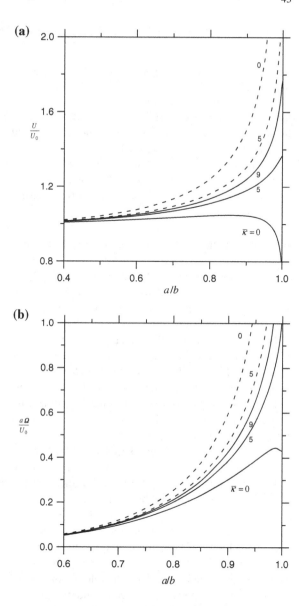

both are the same. The rotation velocity of regularization osmophoresis $\Omega/U_0$ is approximate to the trend of increase or decrease of $U/U_0$ on the plate of solute impermeable, as the value of $a\Omega/U_0$ reduces gradually with increase of parameter $\kappa$ and decrease of parameter $\bar{\kappa}$; in the situation of linear solution distribution of plates, the rotation velocity of regularization osmophoresis $a\Omega/U_0$ increases gradually with the increase of parameter $\kappa$ and decrease of parameter $\bar{\kappa}$.

## 3.2.1 Osmophoresis of Particle Parallel to Two Plane Walls

Table 3.3 compares when spherical vesicle particles are at the middle of the two plane walls (when $b = c$ and $\Omega = 0$), osmohporesis is between two parallel plane walls for different coefficients $k$, $\bar{\kappa}$ and $a/b$ is under two different boundary conditions. The results obtained by boundary collocation method are compared with the approximate results obtained by reflection method for mutual verification (see formula [C2-20] in Appendix C).

The results are approximate to the motion of vesicle particle parallel to single plane wall, in the results obtained by boundary collocation method and reflection method (formula C2-20), while in $\lambda \leq 0.6$, the results obtained by both methods match each other closely. When $\lambda \geq 0.8$, there are considerable errors in results obtained by reflection method. Generally speaking, formula (C2-20) underestimates the velocity of osmophoresis of vesicle particles. In comparison with Tables 3.1, 3.2, and 3.3 we realize that when $a/b$ is smaller, the boundary effects of single plate are added directly into it, the boundary effect of the two plane walls are underestimated. However, when $a/b$ is large, the boundary effect of a single plate is directly added into it, and the boundary effect of the two plane walls are overestimated.

For vesicle particles of different coefficients $k$, $\bar{\kappa}$ and $a/b$ located at the middle of two plane walls, the moving velocity of regularization osmophoresis $U/U_0$ is shown as in Fig. 3.4. It is approximate to the results of single plate: on the plates of solute impermeable, as the value of $U/U_0$ reduces gradually with increase of parameter $\kappa$ and decrease of parameter $\bar{\kappa}$; in situation of linear solution distribution of plates, the rotation velocity of regularization osmophoresis $U/U_0$ increases gradually with increase of parameter $\kappa$ and decrease of parameter $\bar{\kappa}$.

We can see from the figures that when the plane walls are solute impermeable and $\kappa \gg 1 + \bar{\kappa}$, or the plane walls are in linear concentration distribution and $\kappa \ll 1 + \bar{\kappa}$, the velocity of regularization osmophoresis $U/U_0$ is started from $a/b = 0$, and is maximized by the increase of $a/b$, and it is gradually decreased by then, even lower than the value it was when $a/b = 0$. This is due to the results from reduction effect of velocity of solute concentration larger than acceleration effect of velocity of fluid force, and the results match with formula (C2-20).

Comparing Figs. 3.4 and 3.2a and 3.3a shows that, when adding a second plate, it may not enhance its influence on velocity of osmophoresis of particles (even the equal distance of the two plane walls and particle). When adding the second plate, although the acceleration effect and solute concentration gradient are all enhanced, the level of which varies, its total influence may not enhance the velocity of osmophoresis of vesicle particles.

The vesicle particles located in any position between two plane walls, for different separation parameters $a/b$, when $\kappa = 1 + \bar{\kappa}$ (There are the same numerical results of two different boundary conditions), the numerical results of moving velocity $U/U_0$ and rotation velocity $a\Omega/U_0$ of regularization osmophoresis of vesicle particles are shown in Fig. 3.5, wherein, the dotted line represents when

**Table 3.3** Normalized osmophoretic velocity of a spherical vesicle along the median plane between two parallel plane walls

| a/b | $\kappa = \bar{\kappa} = 0$ | | $\kappa = 10\ \bar{\kappa} = 0$ | | $\kappa = 0\ \bar{\kappa} = 10$ | |
|---|---|---|---|---|---|---|
| | Exact solution | Asymptotic solution | Exact solution | Asymptotic solution | Exact solution | Asymptotic solution |
| *For impermeable plane walls* | | | | | | |
| 0.2 | 1.00769 | 1.00768 | 1.00466 | 1.00466 | 1.00769 | 1.00768 |
| 0.4 | 1.05744 | 1.05679 | 1.03219 | 1.03158 | 1.05752 | 1.05679 |
| 0.6 | 1.17967 | 1.16731 | 1.08410 | 1.07284 | 1.18171 | 1.16731 |
| 0.8 | 1.41975 | 1.32204 | 1.12970 | 1.05479 | 1.44544 | 1.32204 |
| 0.9 | 1.65366 | 1.39811 | 1.11592 | 0.97239 | 1.75945 | 1.39811 |
| 0.95 | 1.8903 | 1.43086 | 1.0659 | 0.89872 | 2.1753 | 1.43086 |
| 0.99 | 2.5380 | 1.45290 | 0.8984 | 0.81936 | 3.9278 | 1.45290 |
| 0.995 | 2.763 | | 0.833 | | 4.690 | |
| 0.999 | 2.949 | | 0.727 | | 5.392 | |
| *For plane walls prescribed with the far-field solute concentration profile* | | | | | | |
| 0.2 | 1.00557 | 1.00556 | 1.00784 | 1.00784 | 1.00557 | 1.00556 |
| 0.4 | 1.03967 | 1.03894 | 1.05880 | 1.05837 | 1.03959 | 1.03894 |
| 0.6 | 1.11187 | 1.09879 | 1.18650 | 1.17561 | 1.11008 | 1.09879 |
| 0.8 | 1.21302 | 1.12156 | 1.45732 | 1.35552 | 1.19382 | 1.12156 |
| 0.9 | 1.27225 | 1.07292 | 1.76466 | 1.46018 | 1.21050 | 1.07292 |
| 0.95 | 1.3081 | 1.02076 | 2.1223 | 1.51388 | 1.1851 | 1.02076 |
| 0.99 | 1.3557 | 0.96126 | 3.1040 | 1.55682 | 1.0744 | 0.96126 |
| 0.995 | 1.349 | | 3.352 | | 1.023 | |
| 0.999 | 1.283 | | 3.400 | | 0.920 | |

the fixed distance between a plate with a vesicle particle ($a/b$ = constant), the influence of changing another plate (located at $z = c$) on osmophoretic motion of vesicle particles. The solid line represents when the fixed gap between two plates (when $2a/(b + c)$ = constant), the influence of different location of particles within the gap osmophoretic motion of vesicle particles. As shown in Fig. 3.5a, in this given situation, the net effect of plates will increase the velocity of osmophoresis of particles $U/U_0$. In the situation of a fixed value of $2a/(b + c)$, when vesicle particles locating at the middle of both plates (when $c = b$), it has the minimum moving velocity (its rotation velocity is zero). When vesicle particles are gradually approaching one of the plates ($b/(b + c)$ becoming smaller), the moving and rotation velocity increases gradually.

When the distance between vesicle particles and a plate is fixed (when $a/b$ is fixed), the presence of another plate will increase moving velocity of particles. The closer the distance is between particles and a plate (when $b/(b + c)$ increasing gradually), the rotation velocity of particles will gradually decrease.

We have found that when two plane walls are solute impermeable and $\kappa \gg 1 + \bar{\kappa}$, or two plates are linear solute distribution and $\kappa \ll 1 + \bar{\kappa}$, the net effect of plates will be able to reduce velocity of osmophoresis of particles.

**Fig. 3.4  a** When $\bar{\kappa} = 10$, the velocity of osmophoresis of $U/U_0$ to $a/b$ of vesicle particles parallel to two plane walls (when $c = b$), **b** The osmophoretic velocity of $U/U_0$ to $a/b$ of vesicle particle parallel to the two plane walls (when $c = b$) when $\kappa = 10$

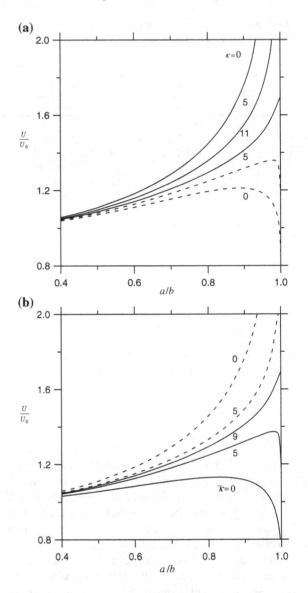

By cross-referencing several different cases of $2a/(b + c)$, we find when the vesicle particle locates at the middle of two plane walls, there is a relative minimum value of velocity of vesicle particles; when the vesicle particles approach a plate, its velocity is increasing. What is worth mentioning is that with a fixed distance between vesicle particles and a plate (a fixed value of $a/b$), the influence of another plate on motion velocity of vesicle particles is not just monotonic functions, however, for the sake of brevity, this is no longer illustration for figures.

**Fig. 3.5** **a** When $\kappa = 1 + \bar{\kappa}$, the moving velocity of osmophoresis of $U/U_0$ to $b/(b + c)$ of vesicle particles parallel to two plane walls (the *solid line* represents the fixed $2a/(b + c)$, while the *dotted line* is the fixed $a/b$), **b** when $\kappa = 1+(\bar{\kappa})$, the rotation velocity of osmophoresis of $a\Omega/U_0$ to $b/(b + c)$ of vesicle particles parallel to two plane walls

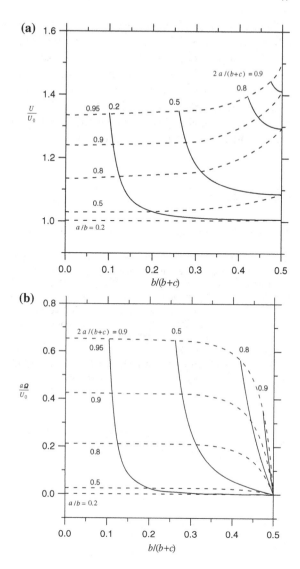

## 3.3 Conclusions

This study considers a single spherical colloidal particle in the case of the low Reynolds number and low pictogram number of columns, parallel in a single infinite plate or infinite plate of osmophoretic motion behavior. Respectively, taking point method (boundary collocation method) with the reflection method for solving particle swimming velocity and comparing particle phoretic motion in the case.

In this chapter, I calculated single spherical vesicle particles swimming moving velocity of penetration of the solution degrees. The boundary conditions of the plate cannot penetrate the solute and linear distribution of the two situations. The velocity of the semipermeable membrane separation parameters ($=a/b$) is a monotonically increasing function. However when $a/b$ is closing to 1, in the case of the different parameters $k$ and $\bar{\kappa}$, the boundary of the plate efficiency should be the velocity of moving that can accelerate or reduce vesicle particles (with a single vesicle particle in the boundary condition).

# References

Anderson, J.L.: Movement of a semipermeable vesicle through an osmotic gradient. Phys. Fluids **26**, 2871 (1983)

Ganatos, P., Weinbaum, S., Pfeffer, R.: A strong interaction theory for the creeping motion of a sphere between plane parallel boundaries. Part 2. Parallel motion. J. Fluid Mech. **99**, 755 (1980)

Keh, H.J., Jan, J.S.: Boundary effects on diffusiophoresis and electrophoresis: Motion of a colloidal sphere normal to a plane wall. J. Colloid Interface Sci. **183**, 458 (1996)

Keh, H.J., Yang, F.R.: Boundary effects on osmophoresis: motion of a vesicle normal to a plane wall. Chem. Eng. Sci. **48**, 609 (1993a)

Keh, H.J., Yang, F.R.: Boundary effects on osmophoresis: motion of a vesicle in an arbitrary direction with respect to a plane wall. Chem. Eng. Sci. **48**, 3555 (1993b)

O'Brien, V.: Form factors for deformed spheroids in Stokes flow. AIChE J. **14**, 870 (1968)

# Chapter 4
# The Thermocapillary Motion of Spherical Droplet Parallel to the Plane Walls

**Abstract** The thermocapillary motion velocity of single spherical droplet will be calculated ignoring the flow inertia and the temperature. Applying the value of temperature gradient to that is parallel to the flat plate will be its driving force, and assuming that the droplets will be kept doing spherical motion without deformation. The boundary conditions of the plate can be discussed in two situations: the linear distributions of adiabatic and temperature. When the liquid drops approach to the plate, one of the boundary effects of the plate comes from the gradient interaction of the temperature between the droplet and the plate, and the other comes from the viscosity effect of the fluid. This chapter uses the boundary collocation method to calculate the fluid in the thermocapillary of the viscosity ratio, thermal conductivity ratio as well as the separation parameters velocity, and comparing the calculated results from the reflection method. Their results are consistent. As for the plate boundary effect, due to the characteristics of the droplets, the relative distance of the droplet with the flat plate, and the boundary conditions of the different plane walls, the fluid droplet velocity can be increased or reduced.

## 4.1 Theoretical Analysis

This chapter considers the radius of a spherical droplet, affected by the applied temperature field, carried out parallel to the two flat steady state thermocapillary moving. The center of the spherical droplets, from the two-plate distance b and c, as shown in Fig. 4.1. The wherein $(x, y, z)$, $(p, \emptyset, z)$ and $(r, \theta, \varphi)$ represent the spherical droplets central point of origin of cartesian coordinates, cylindrical coordinates, and spherical coordinate systems, respectively.

At infinity, the temperature $T_\infty(x)$ performs a linear distribution, the temperature gradient is $E_\infty e_x (= \nabla T_\infty)$, and $e_x$ is the unit vector in the direction of the Cartesian coordinates. Assuming capillary number $\eta U_0/\gamma$ is sufficiently small (wherein $U_0 = |U_0|$, see formula (1.8)), interfacial tension is sufficient to enable

P.-Y. Chen, *The Application of Biofluid Mechanics*, SpringerBriefs in Physics, DOI: 10.1007/978-3-642-44952-9_4, © The Author(s) 2014

**Fig. 4.1** The coordinates graph of the thermocapillary motion of a single spherical droplet parallel to two plates

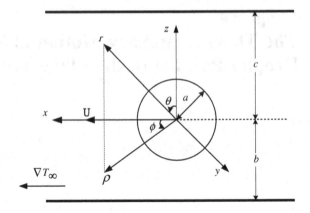

droplet during thermocapillary motion while maintaining the spherical gravitational convection term effect can be ignored. This will be theoretical calculations estimated with the presence of formula (1.8) represented by the separate droplets thermocapillary the amendments made to thermocapillary velocity. Before solving the droplet internal and external fluid velocity field, you must first obtain the distribution of droplet temperature in internal and external fields.

### 4.1.1 Temperature Distribution

Here, considering the Peclet number is quite small and can be ignored. Hence, in the outer region, Laplace equation can be the major equation for the temperature field $T(x)$:

$$\nabla^2 T = 0 \quad (r \geq a) \tag{4.1}$$

while the main equations for the temperature field $T_1(x)$ within the droplets is:

$$\nabla^2 T_1 = 0 \quad (r \leq a)$$

They meet the boundary conditions, in order to meet the droplets internal and external to the temperature distribution, and the droplet surface method, the line direction of the temperature gradient must be continuous and meet the condition of far-field temperature of the droplet without disturbance, thus:

$$r = a: \quad T = T_1; \quad k\frac{\partial T}{\partial r} = k_1\frac{\partial T_1}{\partial r} \tag{4.2a, b}$$

$$z = c, -b: \quad \frac{\partial T}{\partial z} = 0 \tag{4.2c}$$

$$\rho \to \infty: \quad T = T_0 + E_\infty x \tag{4.2d}$$

Wherein, k and $k_1$ as the external fluid and the droplets of thermal conductivity, and $T_0$ is the droplet center position when droplets do not exist the temperature. Wherein formula (4.2c) for two adiabatic plates boundary conditions. For linear distribution of temperature, the boundary conditions of two plates, formula (4.2c) should be revised to:

$$z = c, -b: \quad T = T_0 + E_\infty x \tag{4.2e}$$

In the special case of $k_1 = k$, using formula (4.2c) and (4.2e) to calculate the temperature field, there will be the same results.

As the governing equation and boundary conditions are linear, so the temperature distribution $T$ can be expressed as below:

$$T = T_w + T_s \tag{4.3}$$

Wherein, $T_w$ is the presence of plates disturbed by formula (4.1), double fourier integral general solution labeled by Cartesian coordinates, plus with far-field temperature distribution without disturbance, which can be performed as below:

$$T_w = T_0 + E_\infty x + E_\infty \int_0^\infty \int_0^\infty (X e^{\kappa z} + Y e^{-\kappa z}) \sin(\alpha x) \cos(\beta y) d\alpha \, d\beta \tag{4.4}$$

Wherein, $X$ and $Y$ are undetermined functions; and $\kappa = (\alpha^2 + \beta^2)^{1/2}$. As for $T_s$, satisfying formula (4.1), is spherical coordinates general solution caused by disturbance due to the presence of droplets, which is also a spherical harmonic function:

$$T_s = E_\infty \sum_{n=1}^\infty R_n r^{-n-1} P_n^1(\mu) \cos \varphi \tag{4.5}$$

Wherein, $P_n^1$ is accompanied by associated Legendre function, while $\mu$ representing $\cos\theta$ in order to be concise, and $R_n$ is unknown coefficient. Using the temperature distribution T shown in formula (4.3–4.5) is already satisfying the boundary conditions of the infinity formula (4.2d). The general solution of temperature field within the droplet can be performed as:

$$T_1 = T_0 + E_\infty \sum_{n=1}^\infty \overline{R}_n r^n P_n^1(\mu) \cos \varphi \tag{4.6}$$

Wherein, $\overline{R}_n$ is unknown constants.

Herein, we introduce process of solving the unknown coefficients $X$, $Y$, $R_n$, and $\overline{R}_n$. Applying the temperature distribution $T$ in formula (4.3–4.6) into boundary conditions of formula (4.2c) (or 4.2e), after processing Fourier sine and cosine transforms of x and y, they can be performed as $R_n$ and $\overline{R}_n$ in equations. Then, substituting the general solution of them into formula (4.4) can perform the integral types of the temperature distribution T as the modified Bessel functions of the second kind as below:

$$T = T_0 + E_\infty x + E_\infty \sum_{n=1}^{\infty} R_n \delta_n^{(1)}(r, \mu) \cos \varphi \qquad (4.7)$$

Wherein, the details of equations $\delta_n^{(1)}(r, \mu)$ please refer to formula (D1) in Appendix D. Substituting formulas (4.6) and (4.7) into boundary conditions formula (4.2a,b), we can obtain:

$$\sum_{n=1}^{\infty} [R_n \delta_n^{(1)}(a, \mu) - \overline{R}_n a^n P_n^1(\mu)] = -a(1 - \mu^2)^{1/2} \qquad (4.8a)$$

$$\sum_{n=1}^{\infty} [R_n \delta_n^{(2)}(a, \mu) - \overline{R}_n k^* n a^{n-1} P_n^1(\mu)] = -(1 - \mu^2)^{1/2} \qquad (4.8b)$$

Wherein, the details of equations $\delta_n^{(2)}(r, \mu)$ please refer to formula (D2), and $k^* = k_1/k$. And its integral $\delta_n^{(1)}$ and $\delta_n^{(2)}$ are obtained via numerical integration.

For each droplet surface, it will require infinite undetermined coefficients, $R_n$ and $\overline{R}_n$, to comply with the boundary conditions of formula (4.8a, b). However, we can abridge infinite series into limited series by using boundary collocation method, and take a limited number of points on the surface of each droplet to satisfy the boundary conditions. And please refer to Chap. 2 for the mathematical approach for it.

### 4.1.2 Distribution of Fluid Velocity

The temperature distribution obtained in previous section can be utilized to calculate the distribution of fluid velocity in the system. Assuming the internal and external fluid of droplets are incompressible Newtonian fluid. Due to thermocapillary motions are as creeping flows, the internal and external flow fields of droplets can be performed with Stroke equations as below:

$$\eta \nabla^2 v - \nabla p = 0; \quad \nabla \cdot v = 0 \quad (r \geq a) \qquad (4.9a, b)$$

$$\eta_1 \nabla^2 v_1 - \nabla p_1 = 0; \quad \nabla \cdot v_1 = 0 \quad (r \leq a) \qquad (4.10a, b)$$

Wherein, $v_1(x)$ and $v(x)$ are the velocity distribution of the internal and external flow fields, respectively, and p represents stress distribution.

On the droplet surface, the boundary conditions of fluid on plates and infinity are (Young et al. 1959; Anderson 1985):

$$r = a: \quad e_r \cdot (v - U) = 0 \quad v = v_1 \qquad (4.11a, b)$$

$$(I - e_r e_r)e_r : (\tau - \tau_1) = -\frac{\partial \gamma}{\partial T}(I - e_r e_r) \cdot \nabla T \qquad (4.11c)$$

$$z = c, -b: \quad v = 0 \tag{4.11d}$$

$$\rho \to \infty: \quad v = 0 \tag{4.11e}$$

Wherein, $\gamma(\theta, \varphi)$ is the value of local interfacial tension of droplets; $\tau = \eta[\nabla v + (\nabla v)^T]$ and $\tau_1 = \eta_1[\nabla v_1 + (\nabla v_1)^T]$ are the internal and external viscous tress tensors of droplets, respectively; $e_r, e_\theta$ and $e_\varphi$ are the unit vector of spherical coordinates, and I is unit tensor; as for $U = Ue_x$ is the velocity thermocapillary motions of droplets. Here, it is assumed that $\partial\gamma/\partial T$ is a fixed value, and $\nabla T$ can be obtained by value of coefficient solved by temperature general solution of formula (4.7) in combination with formula (4.8a, b). While handling asymmetrical problem $b \neq c$, with ignorance of inertial effects, it can be rationally assumed that volecity $U$ and temperature gradient $\nabla T_\infty$ are in the same direction.

The fluid velocity compliant with Stokes equations (formula (4.9)) and the boundary conditions (4.11c) and (4.11d). The general solution of it can be performed as:

$$v = v_x e_x + v_y e_y + v_z e_z \tag{4.12}$$

And,

$$v_x = \sum_{n=1}^{\infty} [A_n(A_n' + \alpha_n') + B_n(B_n' + \beta_n') + C_n(C_n' + \gamma_n')] \tag{4.13a}$$

$$v_y = \sum_{n=1}^{\infty} [A_n(A_n'' + \alpha_n'') + B_n(B_n'' + \beta_n'') + C_n(C_n'' + \gamma_n'')] \tag{4.13b}$$

$$v_z = \sum_{n=1}^{\infty} [A_n(A_n''' + \alpha_n''') + B_n(B_n''' + \beta_n''') + C_n(C_n''' + \gamma_n''')] \tag{4.13c}$$

Among which, the labels as $A_n$, $B_n$, $C_n$, $\alpha_n$, $\beta_n$ and $\gamma_n$ all contain associated Legendre function using $\mu$ (or $\cos\theta$) as even number, and the detailed definition of it please refer to formula (2.6) and the formula (C1) in Ganatos et al. (1980) (see the Appendix D of this paper), and $A_n$, $B_n$, and $C_n$ are undetermined coefficients.

The general solution satisfying the internal flow field of droplets is as below:

$$v_1 = v_{1r} e_r + v_{1\theta} e_\theta + v_{1\varphi} e_\varphi \tag{4.14}$$

Herein,

$$v_{1r} = \sum_{n=1}^{\infty} n P_n^1(\mu)(\overline{C}_n r^{n-1} + \overline{A}_n r^{n+1}) \cos\varphi \tag{4.15a}$$

$$v_{1\theta} = \sum_{n=1}^{\infty} [\bar{B}_n r^n P_n^1(\mu)(1-\mu^2)^{-1/2} - (1-\mu^2)^{1/2} \frac{dP_n^1(\mu)}{d\mu} (\bar{C}_n r^{n-1}$$
$$- \bar{A}_n \frac{n+3}{n+1} r^{n+1})] \cos \varphi \tag{4.15b}$$

$$v_{1\varphi} = \sum_{n=1}^{\infty} [\bar{B}_n r^n (1-\mu^2)^{1/2} \frac{dP_n^1(\mu)}{d\mu} - (1-\mu^2)^{-1/2} P_n^1(\mu)(\bar{C}_n r^{n-1}$$
$$- \bar{A}_n \frac{n+3}{n+1} r^{n+1})] \sin \varphi \tag{4.15c}$$

And $\bar{A}_n$, $\bar{B}_n$ and $\bar{C}_n$ are all unknown coefficients. The general solution is to meet the distribution of flow volecity in any internal location of droplets as conditionality of limited values.

In order to satisfy the boundary conditions of it, substituting formula (4.7) and formulas (4.12–4.15a) into formula (4.11a, b), the obtained results as below:

$$\sum_{n=1}^{\infty} [A_n A_n^*(a, \mu, \varphi) + B_n B_n^*(a, \mu, \varphi) + C_n C_n^*(a, \mu, \varphi)] = U(1-\mu^2)^{1/2} \cos \varphi \tag{4.16a}$$

$$\sum_{n=1}^{\infty} [A_n A_n^*(a, \mu, \varphi) + B_n B_n^*(a, \mu, \varphi) + C_n C_n^*(a, \mu, \varphi)] - \sum_{n=1}^{\infty} [\bar{C}_n na^{n-1} P_n^1(\mu)$$
$$+ \bar{A}_n na^{n+1} P_n^1(\mu)] \cos \varphi = 0 \tag{4.16b}$$

$$\sum_{n=1}^{\infty} [A_n A_n^{**}(a, \mu, \varphi) + B_n B_n^{**}(a, \mu, \varphi) + C_n C_n^{**}(a, \mu, \varphi)] - \sum_{n=1}^{\infty} [\bar{B}_n a^n (1-\mu^2)^{-1/2} P_n^1(\mu)$$
$$- \bar{C}_n a^{n-1}(1-\mu^2)^{1/2} \frac{dP_n^1}{d\mu} - \bar{A}_n \frac{n+3}{n+1} a^{n+1}(1-\mu^2)^{1/2} \frac{dP_n^1}{d\mu}] \cos \varphi = 0 \tag{4.16c}$$

$$\sum_{n=1}^{\infty} [A_n A_n^{***}(a, \mu, \varphi) + B_n B_n^{***}(a, \mu, \varphi) + C_n C_n^{***}(a, \mu, \varphi)] - \sum_{n=1}^{\infty} [\bar{B}_n a^n (1-\mu^2)^{1/2} \frac{dP_n^1}{d\mu}$$
$$- \bar{C}_n a^{n-1}(1-\mu^2)^{-1/2} P_n^1(\mu) - \bar{A}_n \frac{n+3}{n+1} a^{n+1}(1-\mu^2)^{-1/2} P_n^1(\mu)] \sin \varphi = 0 \tag{4.16d}$$

$$\sum_{n=1}^{\infty} \{(\frac{\partial}{\partial r} - \frac{1}{r})[A_n A_n^{**}(r, \mu, \varphi) + B_n B_n^{**}(r, \mu, \varphi) + C_n C_n^{**}(r, \mu, \varphi)]$$

$$-\frac{(1-\mu^2)^{1/2}}{r} \frac{\partial}{\partial \mu}[A_n A_n^*(r, \mu, \varphi) + B_n B_n^*(r, \mu, \varphi) + C_n C_n^*(r, \mu, \varphi)]\}_{r=a}$$

$$-\eta^* \sum_{n=1}^{\infty}[\overline{B}_n(n-1)a^{n-1}P_n^1(\mu)(1-\mu^2)^{-1/2}$$

$$-\overline{C}_n 2(n-1)a^{n-2}(1-\mu^2)^{1/2}\frac{dP_n^1(\mu)}{d\mu}$$

$$-\overline{A}_n \frac{n(n+2)}{n+1}a^n(1-\mu^2)^{1/2}\frac{dP_n^1(\mu)}{d\mu}]\cos\varphi$$

$$= -\frac{E_\infty}{\eta}\frac{\partial\gamma}{\partial T}[a\mu + \sum_{n=1}^{\infty} R_n \delta_n^{(3)}(a, \mu)]\cos\varphi \qquad (4.16e)$$

$$\sum_{n=1}^{\infty}\{(\frac{\partial}{\partial r} - \frac{1}{r})[A_n A_n^{***}(r, \mu, \varphi) + B_n B_n^{***}(r, \mu, \varphi) + C_n C_n^{***}(r, \mu, \varphi)]$$

$$+\frac{(1-\mu^2)^{-1/2}}{r}\frac{\partial}{\partial\varphi}[A_n A_n^*(r, \mu, \varphi) + B_n B_n^*(r, \mu, \varphi) + C_n C_n^*(r, \mu, \varphi)]\}_{r=a}$$

$$-\eta^* \sum_{n=1}^{\infty}[\overline{B}_n(n-1)a^{n-1}(1-\mu^2)^{1/2}\frac{dP_n^1(\mu)}{d\mu} - \overline{C}_n 2(n-1)a^{n-2}(1-\mu^2)^{-1/2}P_n^1(\mu)$$

$$-\overline{A}_n \frac{n(n+2)}{n+1}a^n(1-\mu^2)^{-1/2}P_n^1(\mu)]\sin\varphi$$

$$= \frac{E_\infty}{\eta}\frac{\partial\gamma}{\partial T}[-a(1-\mu^2)^{1/2} - \sum_{n=1}^{\infty} R_n \delta_n^{(1)}(a, \mu)]\sin\varphi$$

$$(4.16f)$$

Wherein, the detailed definition of function $\delta_n^{(3)}(r, \mu)$ in formula (4.16e) please refer to formula (D3) in Appendix D, as for the functions with superscripted asterisk, An, Bn, and Cn, about those definition refer to (D7), and $\eta^* = \eta_1/\eta$. As for the first M item of coefficient $R_n$ can be decided by method in the previous section.

After detailed observation on formula (4.16a), we can discover that: While using the boundary collocation method at the boundary of the ball surface, $r = a$, all the associated vertical are not related to the selection of the value of $\varphi$. Therefore, the formula (4.16a) is compliant with N of different values of $\theta_i$ on each surface of spheric particle (values of $\theta$ are between 0 and $\pi$) resulting in 6N of linear equation, and just can be applied to solve 6N of unknowns, $A_n, B_n, C_n, \overline{A}_n, \overline{B}_n$ and $\overline{C}_n$, and the distribution of fluid field could be obtained smoothly when N is large enough.

### 4.1.3 Deduction of Droplet Thermocapillary Velocity

The drag force is applied to the spherical droplets by fluid can be performed as (Ganatos et al. 1980):

$$F = -8\pi\eta A_1 e_x \tag{4.17}$$

From the above equation: we can know that only low-order coefficient $A_1$ had contributed to the drag force applied to the spherical droplets.

Droplets are freely suspended in solution, the net force received by them is zero. Using this limitation in formula (4.17), we can obtain:

$$A_1 = 0 \tag{4.18}$$

In combination with formula (4.18) and 6N linear equations generated by the formula (4.16a), the moving velocity U of droplets can be smoothly obtained.

## 4.2  Results and Discussion

This section will explore the use of the boundary collocation method for solving the calculated results of thermocapillary motion of single spherical droplet parallel to two plates. As for the calculation of the boundary collocation method, please refer to Chap. 2.

### 4.2.1  The Thermocapillary Motion of Spherical Droplet Parallel to Single Plate

In Table 4.1, respectively comparing the resulted numeric values of different parameters, $k^*$, $\eta^*$, and a/b, by the boundary boundary collocation methods, under two different plates boundary conditions, through the thermocapillary motion of spherical droplets parallel to single plate (when $c \to \infty$), and processing mutual verification with approximate results generated from reflection method.

In Table 4.1, both using the numerical calculation of the boundary collocation method converges to present effective digits. Convergence rate is related to values of a/b, the lager the values are, the slower the convergence rate is, and also the more needed points to be accessed. When a/b = 0.999, the boundary collocation method must reach $M = 36$ and $N = 36$, or above, to be able to start convergence.

In Appendix C3, I illustrate the spherical droplet, parallel to plates, with reflection method and obtain approximate analytical solution of thermocapillary motion in details, and the moving and rotation velocity of droplets can be found in formula (C3–11), the results of this calculation are also listed in the table, with the

**Table 4.1** The thermocapillary motion of spherical droplet parallel to single insulated plate

| $a/b$ | $U/U_0$ | | | |
|---|---|---|---|---|
| | $k^* = \eta^* = 0$ | | $k^* = \eta^* = 10$ | |
| | Exact solution | Asymptotic solution | Exact solution | Asymptotic solution |
| *For an insulated plane wall* | | | | |
| 0.2 | 0.99950 | 0.99950 | 0.99828 | 0.99827 |
| 0.4 | 0.99577 | 0.99576 | 0.98648 | 0.98637 |
| 0.6 | 0.98339 | 0.98377 | 0.95378 | 0.95238 |
| 0.8 | 0.94340 | 0.95264 | 0.87831 | 0.87311 |
| 0.9 | 0.88498 | 0.92330 | 0.79943 | 0.80170 |
| 0.95 | 0.8186 | 0.90334 | 0.7269 | 0.75369 |
| 0.99 | 0.6989 | 0.88419 | 0.6155 | 0.70780 |
| 0.995 | 0.675 | | 0.594 | |
| 0.999 | 0.654 | | 0.575 | |
| *For a plane wall prescribed with the far-field temperature profile* | | | | |
| 0.2 | 0.99849 | 0.99849 | 0.99978 | 0.99978 |
| 0.4 | 0.98764 | 0.98763 | 0.99871 | 0.99860 |
| 0.6 | 0.95481 | 0.95531 | 0.99709 | 0.99552 |
| 0.8 | 0.86997 | 0.88045 | 0.99327 | 0.98395 |
| 0.9 | 0.77497 | 0.81557 | 0.98279 | 0.96846 |
| 0.95 | 0.6879 | 0.77320 | 0.9633 | 0.95606 |
| 0.99 | 0.5577 | 0.73348 | 0.9175 | 0.94302 |
| 0.995 | 0.533 | | 0.909 | |
| 0.999 | 0.513 | | 0.904 | |

boundary collocation method to get the correct value results comparable to each other. The results showed that when $\lambda \leq 0.8$, the results of regularization moving velocity obtained by the boundary collocation method agreed well with the numerical values obtained by the reflection method, and the error is less than 1.3 %. However, when the droplet-plate gap is very small (i.e., when $\lambda \geq 0.9$), the accuracy of reflection method is much lower. Generally speaking, when $\lambda$ approaching 1, the velocity of thermocapillary motion of droplets will be over-estimated in formula (C3–11).

For different parameters k*, $\eta^*$, and $a/b$, the moving velocity of thermocapillary motion of droplets $U/U_0$, the numerical results of which is shown in Figs. 4.2 and 4.3. As shown in figures, if other conditions are remained unchanged ($\eta^*$ and $a/b$ remained unchanged) (boundary conditions of the formula 4.2c), the regularization thermocapillary motion velocity in the case of the insulated plane walls, the parameter $U/U_0$ will reduce gradually with the increasing values of parameters $\kappa^*$. However, on the plates of linear temperature distribution (boundary conditions of the formula 4.2e), then the regularization thermocapillary motion velocity $U/U_0$ will increase with the increasing values of parameters $\kappa^*$.

**Fig. 4.2 a** The thermocapillary motion velocity of $U/U_0$ to $a/b$ of droplet parallel to the single plate when $k^* = 0$ (wherein, the *solid line* represents the insulated plate, the *dotted line* represents the linear temperature distributed plate, and the *black dots* are estimated situation of $k^* = \eta^* = 0$ possessed by Meyyappan and Subramanian (1987)). **b** The thermocapillary motion velocity of $U/U_0$ to $a/b$ of droplet parallel to the single plate when $k^* = 100$ (wherein, the *solid line* represents the insulated plate, the *dotted line* represents the linear temperature distributed plate)

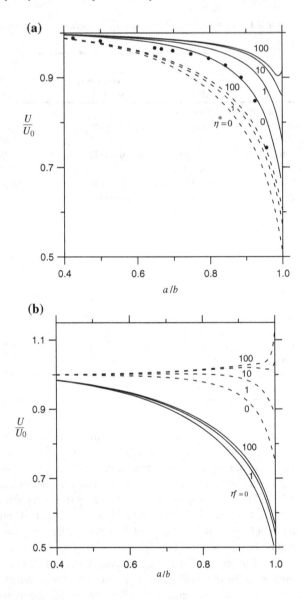

Detailedly observing Figs. 4.2 and 4.3, we will find an interesting phenomenon: when the insulated plate is with large $\eta^*$ and small $k^*$ (i.e., $\eta^* = 100$, $k^* = 0$), and the value of $a/b$ is smaller, the mobility of thermocapillary motion of droplets will decrease gradually to a minimum value with increasing values of $a/b$, and will increase gradually with constantly increasing values of $a/b$ by then. If the droplet-plate gap is small enough, the velocity of its motion will even be greater than it is with the absence of plates. On the contrary (when large $k^*$ and small $\eta^*$), the

**Fig. 4.3** With different values of $\eta^*$, use the mobility of normalized thermocapillary motion (when $k^* = 1$, *solid lines*) to make the figure of mobility of sedimentation and $a/b$ (*dotted lines*)

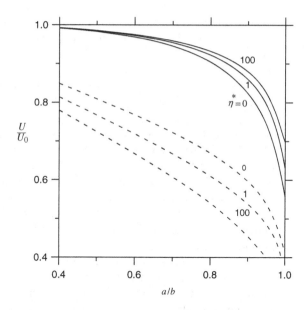

velocity of its motion will decrease monotonically and gradually with increasing values of $a/b$.

For the plate is linear temperature distribution, while in the case of large $\eta^*$ and small $k^*$ (i.e., $\eta^* = 100$, $k^* = 0$), and the value of a/b is smaller, the mobility of thermocapillary motion of droplets will decrease gradually to a minimum value with increasing values of $a/b$, and will increase gradually with constantly increasing values of a/b by then. And if the droplet-plate gap is small enough, the velocity of its motion will even be greater than it is with the absence of plates. For instance, when $k^* = \eta^* = 100$ and $a/b = 0.999$, the velocity of its motion will be 13 % faster than it is with the absence of plates. On the contrary (when $k^*$ and $\eta^*$ are very small), the velocity of its motion will decrease monotonically and gradually with increasing values of $a/b$.

In cases of certain values of $\eta^*$ (when $\eta^* = 1$–$10$ and $k^* = 100$), the mobility of droplets varying with change of $a/b$ (wherein, $0 < a/b < 1$) will generate 2–3 of extreme values (Max. or Min.).

With gradually increasing of $a/b$, there is non-monotonic change of the mobility of droplets with prior decreasement and latter increasement, and the detailed instruction of this is as below:

(1) In this case, in the droplet thermal conductivity $k_1$ is zero ($k^* = 0$ at the time), insulated plates: under these conditions, the droplet is not thermally conductive, so the heat line does not pass through the droplets, the heat flow lines nor plate absorbed. Accordingly, in the droplets gradually close to the flat plate, and its droplet with the plate between the heat line density is increased (the temperature gradient increase), and therefore on the droplet acceleration effect. Plate caused by the presence of the droplet viscosity effect, it will reduce the

velocity of the droplets. Result of two effects competing with each other, resulting in gradually increasing (when droplet gradually approaching plates), the phenomenon of droplet mobility have prior decreasement and latter increasement.

(2) Droplets of thermal conductivity k is zero ($k^* = 0$ at the time), a plate for the linear temperature distribution is as below: At this point, because the droplet is not thermally conductive, heat flow lines do not pass through the droplet, however, for the plate in condition of linear temperature distribution, heat flow lines can be absorbed by plates. Thus, when the droplet neighboring the plate, the density of heat flow lines between the droplet and plate reduced (temperature gradient reduced), and thus, it has the effect of reduction of velocity of the droplet. And the presence of the plate resulted in viscous drag effects to the droplet, and will decelerate its velocity as well. Therefore, when a/b is gradually increasing, the mobility of the droplet rendered monotonically and gradually decreasing.

(3) The thermal conductivity of the droplet $k_1 \gg k$ ($k^*$ is very large at the time), with insulated plates: under this conditions, since the droplets are very easy to thermal conductivity, so the heat flow lines can be absorbed easily by droplets, therefore, when droplets approaching insulated plates, the density of heat flow lines between the droplets and plates will be reduced (temperature gradient is reduced), and thus is have effects of reduction of velocity of droplets. However, the presence of the plate resulted in viscous drag effects to the droplet, and will decelerate its velocity as well. Therefore, when a/b is gradually increasing, the mobility of the droplet rendered a decreasing trend monotonically and gradually.

(4) The thermal conductivity of the droplets $k_1 \gg k$ ($k^*$ is very large at the time), with linear temperature distributed plates, At the moment, the droplets is very easy to thermal conductivity, and heat flow lines will be absorbed very easily. Thus, when droplets neighboring linear temperature distributed plates, the density of heat flow lines between the droplets and plates will be increased (temperature gradient increased), which have effects of acceleration of droplets. However, the presence of the plate resulted in viscous drag effects to the droplet, and will decelerate its velocity as well. Thus, when a/b is gradually increasing, the phenomenon of droplet mobility have prior decreasement and latter increasement.

(5) When $k^* = 1$, under boundary conditions of two different plate, there will be the same effect on thermocapillary motion of droplets. Under this special condition, the temperature interaction between the plate and the droplet disappeared. Since only the presence of plates will have viscous drag effects on droplets, thus, the mobility of thermocapillary motion of droplets will decrease monotonically with increasing a/b.

And the result is in consistence with calculation results of the reflection method. (see analysis in Appendix C3).

**Table 4.2** The regularization velocity value $U/U_0$ of thermocapillary motion when single spherical bubble (when $k^* = \eta^* = 0$) parallel to linear temperature distributed plate

| $a/b$ | Meyyappan and Subramanian | Collocation solution | Asymptotic solution |
|---|---|---|---|
| 0.95666 | 0.74314 | 0.67184 | 0.76697 |
| 0.92498 | 0.84781 | 0.73728 | 0.79534 |
| 0.88684 | 0.89967 | 0.79168 | 0.82551 |
| 0.84353 | 0.92698 | 0.83627 | 0.85509 |
| 0.79669 | 0.94236 | 0.87220 | 0.88221 |
| 0.74772 | 0.95217 | 0.90082 | 0.90592 |
| 0.69779 | 0.95953 | 0.92346 | 0.92592 |
| 0.66667 | 0.96354 | 0.93503 | 0.93656 |
| 0.64805 | 0.96581 | 0.94118 | 0.94231 |
| 0.50000 | 0.98172 | 0.97511 | 0.97516 |
| 0.42510 | 0.98808 | 0.98507 | 0.98507 |
| 0.33333 | 0.99392 | 0.99293 | 0.99293 |
| 0.26580 | 0.99681 | 0.99645 | 0.99645 |
| 0.20000 | 0.99860 | 0.99849 | 0.99849 |
| 0.16307 | 0.99922 | 0.99919 | 0.99919 |
| 0.10000 | 0.99982 | 0.99981 | 0.99981 |
| 0.06667 | 0.99995 | 0.99994 | 0.99994 |
| 0.04000 | 0.99999 | 0.99999 | 0.99999 |

Through the use of bipolar coordinates, Meyyappan and Subramanian (1987) obtained some semianalytical-seminumerical solutions for the normalized themocapillary velocity $U/U_0$ of a spherical gas bubble (with $k^* = \eta^* = 0$) migrating parallel to a plane wall prescribed with the far-field temperature distribution. These solutions are also presented in Fig. 4.2a for comparison. It can be seen that the bipolar-coordinate solution for $U/U_0$ of a bubble at a given value of $a/b$ is far greater than our corresponding collocation solution (illustrated by the lowest dashed curve in Fig. 4.2a).

A detailed comparison given by Table 4.2 shows that our collocation solutions agree much better with the method-of-reflection solution given by C3-11 than the bipolar-coordinate solutions do for all values of $a/b$ less than 0.9. It seems quite likely that these bipolar-coordinate solutions are in error. Note that the relative thermocapillary velocity of a gas bubble near a plane wall always decreases as the relative gap thickness decreases (or $a/b$ increases) no matter whether the wall is insulated or prescribed with the far-field temperature distribution.

Appendix A boundary collocation method obtained in the same geometrical conditions, the droplets to be fixed in the body-force field (body-force field, such as gravity field) moving velocity. In Fig. 4.3 the relatively spherical droplets settlement moving (in this case) thermocapillary action subject to boundary effect. Obviously, the thermocapillary effect of moving plate boundary should be much smaller than the droplets to the case when the settlement moving. When increases, the droplet settlement moving by the plate boundary effect gradually increased, and when the droplets thermocapillary mobility trend to be the contrary.

## 4.2.2 The Thermocapillary Motion of Spherical Droplet Parallel to the Plane Walls

Table 4.3 comparing the spherical droplets at the midpoint of the two plane walls (when b = c), parallel to the two plane walls in the thermocapillary motion, for different $k^*$, $\eta^*$, and $a/b$, in the boundary conditions of two different plates, the numerical results obtained by the boundary collocation method will be mutual verified with the approximate results obtained by the reflection method.

Situation similar to the motion of droplet parallel to single plate, the correct numerical results obtained by the boundary collocation method is compared with approximate results obtained by the reflection method for extending $\lambda$ (=$a/b$) to $O(\lambda^7)$ (formula (C3–20)). When $\lambda \leq 0.6$, these two calculations are very consistent, but when $\lambda \geq 0.8$, there is a considerable error of the approximate results obtained by the reflection method. Generally speaking, the thermocapillary motion velocity of droplets will be overestimated in formula (C3–20). Comparing Tables 4.3 and 4.1 with each other, we can find that: when $a/b$ is smaller, directly summing boundary effects of single plate into together will underestimate the boundary effect of two plates, however, when $a/b$ is larger, directly summing the boundary effect of single plate will overestimate boundary effect of two plates.

When spherical droplet is at middle of two plane walls, for moving velocity $U/U_0$ of regularization thermocapillary motion of different $k^*$, $\eta^*$, and $a/b$, the numerical results obtained by the boundary collocation method is shown in Fig. 4.4. When $\eta^*$ and $a/b$ are fixed, thermocapillary motion of droplets parallel to two linear temperature distributed plates, due to larger droplet-plate temperature gradient, the velocity $U/U_0$ will increase with increasing $k^*$; while being parallel to two insulated plates, it will decrease with increasing $k^*$. As for situation of fixed $k^*$ and $a/b$, under two different boundary effects of plates, the velocity $U/U_0$ will increase with increasing $\eta^*$.

Approximately, the thermocapillary motion of droplets parallel to insulated plane walls with a large $\eta^*$ and small $k^*$ situation, or the thermocapillary motion of droplets parallel to linear temperature distributed plates with large $\eta^*$ and $k^*$, with the increasing $a/b$, the mobility of thermocapillary motion will be decreased to a minimum value prior to appear a trend of monotonically and gradually increasing, its mobility will even be greater than that of absence of plates.

It can be observed, when the droplet-plate gap is extremely small, the acceleration of the temperature gradient is greater than resistance of current force, resulting in the acceleration of velocity of droplet motion. And the trend of approximate results (formula (C3–20)) of $U/U_0$ obtained by the reflection method is compliant with that obtained by the boundary collocation method.

Comparing Figs. 4.2 and 4.4 shows that, when added to the second plate, it may not enhance the influence on the thermocapillary velocity of droplet (even if the two plane walls with droplets of equidistant). For example, when the droplet parallel to single plate with linear temperature distribution, in situation of $k^* = 100$, $\eta^* = 10$, and $a/b = 0.8$, its motion velocity is 1.5 % more than the

**Table 4.3** The mobility of thermocapillary motion of spherical droplets between two parallel plane walls

| $a/b$ | $U/U_0$ | | | |
|---|---|---|---|---|
| | $k^* = \eta^* = 0$ | | $k^* = \eta^* = 10$ | |
| | Exact solution | Asymptotic solution | Exact solution | Asymptotic solution |
| *For insulated plane walls* | | | | |
| 0.2 | 0.99786 | 0.99786 | 0.99496 | 0.99496 |
| 0.4 | 0.98269 | 0.98270 | 0.96228 | 0.96267 |
| 0.6 | 0.93986 | 0.94030 | 0.88483 | 0.89247 |
| 0.8 | 0.84515 | 0.85239 | 0.75406 | 0.81225 |
| 0.9 | 0.75426 | 0.78347 | 0.65752 | 0.79535 |
| 0.95 | 0.6755 | 0.74091 | 0.5863 | 0.80043 |
| 0.99 | 0.5571 | 0.70238 | 0.4915 | 0.81391 |
| 0.995 | 0.535 | | 0.475 | |
| 0.999 | 0.517 | | 0.461 | |
| *For plane walls prescribed with the far-field temperature profile* | | | | |
| 0.2 | 0.99576 | 0.99576 | 0.99811 | 0.99811 |
| 0.4 | 0.96621 | 0.96628 | 0.98692 | 0.98714 |
| 0.6 | 0.88639 | 0.88817 | 0.96491 | 0.96893 |
| 0.8 | 0.72600 | 0.74398 | 0.93828 | 0.96546 |
| 0.9 | 0.59298 | 0.64494 | 0.92380 | 0.98347 |
| 0.95 | 0.4954 | 0.58899 | 0.9092 | 1.00108 |
| 0.99 | 0.3746 | 0.54141 | 0.8785 | 1.02050 |
| 0.995 | 0.354 | | 0.875 | |
| 0.999 | 0.338 | | 0.874 | |

status with absence of plates, while between two symmetrical plane walls, and the rest conditions remain, its motion velocity is slightly 2.5 % smaller than the status with absence of plates.

Thus, when the system added a second plate, the resistance to the role of the current force, and temperature gradient although there are degrees of growth role enhanced contingent in varying degrees, the total influence may not be faster than droplets for thermocapillary velocity.

The droplets located between the two plane walls for different parameters of separation $a/b$, when $k^* = \eta^* = 1$, the numerical results of the moving velocity $U/U_0$ of the droplet normalized thermocapillary are shown in Fig. 4.5. The dotted lines represent when there is a fixed droplet-plate distance ($a/b$ = constant), the imposed influence on thermocapillary motion of droplets via positional change of another plate (located at $z = c$); and the solid lines represent when there is a fixed pitch of the two plates ($2a/(b + c)$ = constant), the imposed influence on thermocapillary motion of droplets via different position of droplets within the pitch. Figure 4.5 shows that, in the given situation, the thermocapillary motion velocity $U/U_0$ will be decelerated by net effects of plates. The greatest moving velocity appears in situation when $2a/(b + c)$ is fixed, the droplet is located at middle of the

**Fig. 4.4  a** The thermocapillary motion velocity of $U/U_0$ to $a/b$ of two symmetrical droplets parallel ($b = c$) to two plane walls when $k^* = 0$ (the *solid line* represents the solute impermeable plane wall, while the *dotted line* for the linear distribution solute plane wall). **b** The thermocapillary motion velocity of $U/U_0$ to $a/b$ of two symmetrical droplets parallel ($b = c$) to two plane walls when $k^* = 100$ (the *solid line* represents the solute impermeable plane wall, while the *dotted line* for the linear distribution solute plane wall)

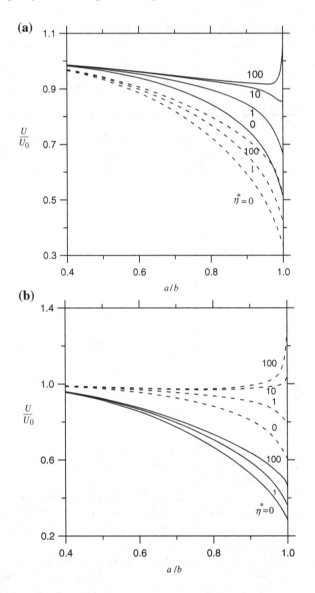

two plates (when $c = b$), and the viscous drag force will be the smallest. When the droplet approaching toward a plate gradually (when $b/(b + c)$ becoming smaller), the drag force upon it will increase to decelerate its velocity.

Under the same geometric conditions, the droplet will be affected by fixed body-force to sedimentation motion, and its moving velocity will be obtained by the boundary collocation method (see in Appendix A). Comparing the sedimentation motion of spherical droplets with influence imposed by boundary effects on thermocapillary motion, it is found that latter is far less than former. When $\eta^*$ is

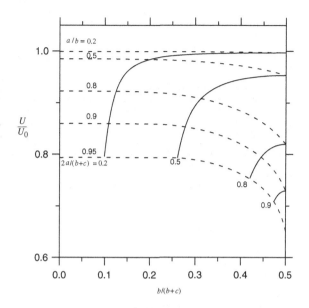

**Fig. 4.5** When $k^* = \eta^* = 1$, the thermocapillary motions velocity of $U/U_0$ to $b/(b + c)$ of droplets parallel to the two plane walls. (the line represents for a fixed $2a/(b + c)$ while the dotted line represents for fixed $a/b$)

increasingly larger, the boundary effects on sedimentation motion of droplets are increasing gradually, and the trend of thermocapillary motion of droplets appears a contrary situation (as shown in Fig. 4.4).

## 4.3 Conclusion

This study considers a single spherical colloidal particles in the case of the low Reynolds number and low picogram number of columns, parallel in a single infinite plate or infinite plate of thermocapillary motion behavior. Respectively, taking point method (boundary collocation method) with the reflection method for solving particle swimming velocity and compare particle phoretic motion in the case.

In this chapter, I calculate a single spherical droplet at a fixed temperature gradient effect of thermocapillary motion. Plate of two kinds of boundary conditions the heat-insulated plate and linear temperature distribution situations are discussed. Overall, the thermocapillary velocity is concerned, showing a monotonically decreasing situation. When the relative viscosity of the droplet $\eta^*$ is large, and $a/b$ tends to 1 for different $k^*$ value of plate boundary effect can increase or reduce the velocity of the droplets of thermocapillary motion. This phenomenon is temperature gradient-enhanced viscous effects and the effect of fluid mechanics of competing with the results.

# References

Anderson, J.L.: Droplet interactions in thermocapillary motion. Int. J. Multiph. Flow **11**, 813 (1985)

Ganatos, P., Weinbaum, S., Pfeffer, R.: A strong interaction theory for the creeping motion of a sphere between plane parallel boundaries. Part 2. Parallel motion. J. Fluid Mech. **99**, 755 (1980)

Meyyappan, M., Subramanian, R.S.: Thermocapillary migration of a gas bubble in an arbitrary direction with respect to a plane surface. J. Colloid Interface Sci. **115**, 206 (1987)

Young, N.O., Goldstein, J.S., Block, M.J.: The motion of bubbles in a vertical temperature gradient. J. Fluid Mech. **6**, 350 (1959)

# Chapter 5
# Thermophoresis Motion of Spherical Aerosol Particles Parallel to Plane Walls

**Abstract** We will discuss the motion velocity of temperature convection when a single spherical aerosol particle is considered ignoring the fluid inertia term of thermophoretic effect. The boundary conditions of the plate can also be divided into two conditions to be discussed, respectively, when the adiabatic temperature displays linear distribution. When the droplets get close to the plate, one of the boundary effects of plate is from the interaction between temperature gradient generated in the aerosol particles and plate, and the other is from the adhesion effect of the fluid stagnation. The boundary-point method is used to obtain the motion velocity of the thermophoretic velocity parameters in different particle thermal conductivity ratios, and particle surface properties of phase related to the separation parameters, when compared with each other and with the reflection method validation. Boundary effect of plane walls may reduce or increase the aerosol particle moving velocity of particles due to the relative distance between the surface characteristics of the aerosol particles, aerosol particles and the plate, and different and boundary conditions of the plate.

## 5.1 Theoretical Analysis

This chapter concerns the situation of the quasi steady thermophoretic motion of particle in a gaseous medium parallel to two plates with the radius of a, which is influenced by external temperature gradient. The distances from the central point of particles to the two plane walls are b and c, respectively, as shown in Fig. 5.1. In this Figure, $(x, y, z)$, $(p, \Phi, z)$, and $(r, \theta, \varphi)$ represent the Cartesian coordinates with the center of particle as the origin, cylindrical coordinates, and spherical coordinate systems. At infinity, the temperature $T_\infty(x)$ displays a linear distribution, and the temperature gradient $-E_\infty e_x$ $(= \nabla T_\infty$, where $E_\infty$ is positive), while $e_x$ is the unit vector of $x$ direction in Cartesian coordinate. We assumed that Knudsen number $1/a$ is small enough to consider the fluid as a continuous body, and the thickness of Knudsen layer is very small in comparison with particle radius or

P.-Y. Chen, *The Application of Biofluid Mechanics*, SpringerBriefs in Physics, DOI: 10.1007/978-3-642-44952-9_5, © The Author(s) 2014

**Fig. 5.1** The
thermophoresis coordinates
figure of single spherical
aerosol particles parallel to
the two plane walls

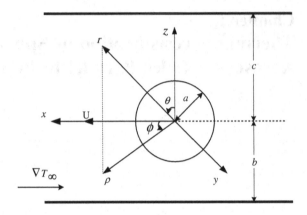

particle–plate pitch. Hence, we will theoretically calculate with the presence of
plates the modification of velocity of thermophoretic motion of single particle
performed by formula (5.1a). And we should obtain the temperature distribution in
internal/external fields prior to solving particle velocity and fluid velocity filed.

### 5.1.1 Distribution of Temperature

In respects to Peclet number is so small that we can ignore it, hence, the Laplace
equation can be used as the main equation of external temperature field $T(x)$:

$$\nabla^2 T = 0 \quad (r \geq a) \tag{5.1a}$$

The main equation of internal temperature filed $T_1(x)$ of particle is:

$$\nabla^2 T_1 = 0 \quad (r \leq a) \tag{5.1b}$$

Boundary conditions are found in the surfaces of the particles, in order to meet
the temperature of the particle surface normal direction of the gradient must be
continuous, and the temperature of the particles inside and outside of the hop
differential is proportional to the normal direction of the temperature gradient, so

$$r = a: \quad k\frac{\partial T}{\partial r} = k_1 \frac{\partial T_1}{\partial r} \tag{5.2a}$$

$$T - T_1 = C_t l \frac{\partial T}{\partial r} \tag{5.2b}$$

where $k$ and $k_1$ are the thermal conductivity of the external fluid and particle,
respectively, and $C_t$ is the hop difference coefficient of particles temperature.

The following is to meet the condition that the temperature away from the
particles is undisturbed:

$$z = c, -b: \quad \frac{\partial T}{\partial z} = 0 \tag{5.2c}$$

$$\rho \to \infty : \quad T = T_0 - E_\infty x \tag{5.2d}$$

where formula (5.2c) is the boundary conditions of the two insulated plates. For the boundary conditions of the linear temperature distribution of the two plates, formula (5.2c) should be revised as:

$$z = c, -b : \quad T = T_0 - E_\infty x \tag{5.2e}$$

Since the governing equations and boundary conditions are linear, the temperature distribution $T$ can be performed as follows:

$$T = T_w + T_p \tag{5.3}$$

where $T_w$ is the temperature presence disturbance caused by formula (5.1a) represented in Cartesian coordinates of the double Fourier integral general solution, plus undisturbed away from the particles at a temperature distribution can be expressed as:

$$T_w = T_0 - E_\infty x - E_\infty \int_0^\infty \int_0^\infty (Xe^{\kappa z} + Ye^{-\kappa z}) \sin(\alpha x) \cos(\beta y) \mathrm{d}\alpha \, \mathrm{d}\beta \tag{5.4}$$

where $X$ and $Y$ are undetermined functions; $\kappa = (\alpha^2 + \beta^2)^{1/2}$. $T_p$, for satisfying the formula (5.1a), the general solution from disturbance caused by the presence of particles in spherical coordinates is a spherical harmonic function:

$$T_p = -E_\infty \sum_{n=1}^\infty R_n r^{-n-1} P_n^1(\mu) \cos \phi \tag{5.5}$$

where $P_n^1$ is associated Legendre function, while $\mu$ represents $\cos\theta$ for simplicity, and $R_n$ stands for the unknown coefficient. The temperature distribution $T$ expressed by formula (5.3)–(5.5) already satisfies the boundary conditions of the infinity of formula (5.2d), as for the general solution of temperature field inside the particle can be performed as:

$$T_1 = T_0 - E_\infty \sum_{n=1}^\infty \overline{R}_n r^n P_n^1(\mu) \cos \phi \tag{5.6}$$

where $\overline{R}_n$ is an unknown constant.

Herein, we introduce process of solving the unknown coefficients $X$, $Y$, $R_n$ and $\overline{R}_n$. The temperature distribution $T$ in formula (5.3)–(5.6) is substituted into boundary conditions of formula (5.2c) (or (5.2e)), and the Fourier sine and cosine transforms of $x$ and $y$ are performed. They can be performed as $R_n$ and $\overline{R}_n$ in equations. Then, substituting the general solution of them into formula (5.4) can perform the integral types of the temperature distribution $T$ as the modified Bessel functions of the second kind as follow:

$$T = T_0 - E_\infty x - E_\infty \sum_{n=1}^{\infty} R_n \, \delta_n^{(1)}(r, \mu) \cos \phi \tag{5.7}$$

where the details of equations $\delta_n^{(1)}(r, \mu)$ please refer to formula [D1] in Appendix D, and substituting formula (5.6) and (5.7) into formulas of boundary conditions (5.2a) and (5.2b), we can obtain:

$$\sum_{n=1}^{\infty} R_n \delta_n^{(2)}(a, \mu) - k^* \sum_{n=1}^{\infty} \overline{R}_n n a^{n-1} P_n^1(\mu) = -(1 - \mu^2)^{1/2} \tag{5.8a}$$

$$\sum_{n=1}^{\infty} R_n [\delta_n^{(1)}(a, \mu) - a C_t^* \delta_n^{(2)}(a, \mu)] - \sum_{n=1}^{\infty} \overline{R}_n a^n P_n^1(\mu) = -a(1 - C_t^*)(1 - \mu^2)^{1/2}$$

$$\tag{5.8b}$$

where $k^* = k_1/k$ and $C_t^* = C_t l/a$. The definition of $\delta_n^{(2)}(r, \mu)$ is in formula [D2]. As for its integral $\delta_n^{(1)}$ and $\delta_n^{(2)}$ are obtained via numerical integration.

For each particle surface, it will require infinite undetermined coefficients, $R_n$ and $\overline{R}_n$, to comply with the boundary conditions of formula (5.8a, 5.8b). However, we can abridge infinite series into limited series by using boundary collocation method, and take a limited number of points on the surface of each particle to satisfy the boundary conditions. And please refer to Chap. 2 for the mathematical approach for it.

### 5.1.2 Distribution of Fluid Velocity

The temperature distribution obtained in Sect. 5.1.1 can be utilized to calculate the distribution of fluid velocity in the system. Assuming the internal and external fluid of particles are incompressible Newtonian fluid. Due to thermophoretic motions are as creeping flows, the internal and external flow fields of particles can be performed with Stroke equations as below:

$$\eta \nabla^2 v - \nabla p = 0 \tag{5.9a}$$

$$\nabla \cdot v = 0 \tag{5.9b}$$

Wherein, $v(x)$ is the velocity distribution of the fluid, and $p$ represents stress distribution of it.

On the particle surface, the boundary conditions of fluid on plates and infinity are (Brock 1962):

$$r = a: \qquad v = U + a\Omega \times e_r + \frac{C_m l}{\eta}(I - e_r e_r)e_r : \tau$$

$$+ C_s \frac{\eta}{\rho \overline{\overline{T}}}(I - e_r e_r) \cdot \nabla T \tag{5.10a}$$

$$z = c, -b: \qquad v = 0 \tag{5.10b}$$

$$\rho \rightarrow \infty: \qquad v = 0 \tag{5.10c}$$

where $\tau = \eta[\nabla v + (\nabla v)^T]$ is the viscous tress tensors of fluid; $e_r$, $e_\theta$ and $e_\phi$ are the unit vector of spherical coordinates, and $I$ is unit tensor; $C_m$ and $C_s$ are friction slip and thermal slip coefficient of particle surface, respectively, and $U = Ue_x$ and $\Omega = \Omega e$ are the moving and rotation velocity of thermophoretic motions of particles. While handling asymmetrical problem ($b \neq c$), by neglecting the inertial effects, it can be rationally assumed that velocity $U$ and temperature gradient $\nabla T_\infty$ are in the same direction.

The velocity compliant with Stokes equations (formulas (5.9a) and (5.9b)) and the boundary conditions (5.10b) and (5.10c). Its general solution can be performed as:

$$v = v_x e_x + v_y e_y + v_z e_z \tag{5.11}$$

where

$$v_x = \sum_{n=1}^{\infty} [A_n(A'_n + \alpha'_n) + B_n(B'_n + \beta'_n) + C_n(C'_n + \gamma'_n)] \tag{5.12a}$$

$$v_y = \sum_{n=1}^{\infty} [A_n(A''_n + \alpha''_n) + B_n(B''_n + \beta''_n) + C_n(C''_n + \gamma''_n)] \tag{5.12b}$$

$$v_z = \sum_{n=1}^{\infty} [A_n(A'''_n + \alpha'''_n) + B_n(B'''_n + \beta'''_n) + C_n(C'''_n + \gamma'''_n)] \tag{5.12c}$$

where the superscripted $\alpha_n$, $\beta_n$, and $\gamma_n$ are all integral form of a function of position (numerical integration), and for its detailed definition please refer to formula [C1] in Ganatos et al. (1980) (see the Appendix D of this book).

In order to satisfy the boundary conditions, by substituting formula (5.7) and (5.11), we can obtain:

$$\sum_{n=1}^{\infty} [A_n A_n^*(a, \mu, \phi) + B_n B_n^*(a, \mu, \phi) + C_n C_n^*(a, \mu, \phi)] = U(1 - \mu^2)^{1/2} \cos \phi$$

$$\tag{5.13a}$$

$$\sum_{n=1}^{\infty} [A_n A_n^{**}(a, \mu, \phi) + B_n B_n^{**}(a, \mu, \phi) + C_n C_n^{**}(a, \mu, \phi)]$$

$$- C_m^* \sum_{n=1}^{\infty} \{(r\frac{\partial}{\partial r} - 1)[A_n A_n^{**}(r, \mu, \phi) + B_n B_n^{**}(r, \mu, \phi) + C_n C_n^{**}(r, \mu, \phi)]$$

$$- (1 - \mu^2)^{1/2} \frac{\partial}{\partial \mu} [A_n A_n^*(r, \mu, \phi) + B_n B_n^*(r, \mu, \phi) + C_n C_n^*(r, \mu, \phi)]\}_{r=a}$$

$$= U\mu \cos\phi + a\Omega \cos\phi - C_s \frac{\eta E_\infty}{\rho T_0} [\mu + \frac{1}{a}\sum_{n=1}^{\infty} R_n \delta_n^{(3)}(a, \mu)] \cos\phi$$

$$(5.13b)$$

$$\sum_{n=1}^{\infty} [A_n A_n^{***}(a, \mu, \phi) + B_n B_n^{***}(a, \mu, \phi) + C_n C_n^{***}(a, \mu, \phi)]$$

$$- C_m^* [\sum_{n=1}^{\infty} \{(r\frac{\partial}{\partial r} - 1)[A_n A_n^{***}(r, \mu, \phi) + B_n B_n^{***}(r, \mu, \phi) + C_n C_n^{***}(r, \mu, \phi)]$$

$$+ (1 - \mu^2)^{-1/2} \frac{\partial}{\partial \phi} [A_n A_n^*(r, \mu, \phi) + B_n B_n^*(r, \mu, \phi) + C_n C_n^*(r, \mu, \phi)]\}_{r=a}$$

$$= - U\sin\phi - a\Omega\mu\sin\phi + C_s \frac{\eta E_\infty}{\rho T_0} [(1 - \mu^2)^{1/2} + \frac{1}{a}\sum_{n=1}^{\infty} R_n \delta_n^{(1)}(a, \mu)] \sin\phi$$

$$(5.13c)$$

where $C_m^* = C_m l/a$; for the definition of $\delta_n^{(3)}(r, \mu)$ please refer to formula [D3] in Appendix D; for the definition of functions with superscripted asterisk $A_n$, $B_n$ and $C_n$ please refer to [D7], and the first $M$ items of coefficient $R_n$ can be determined using the method in Chap. 4.

After detailed observation on formulas (5.13a)–(5.13c), we can discover that: While using the boundary collocation method at the boundary of the ball surface, $r = a$, all the associated vertical are not related to the selection of the value of $\phi$. Therefore, the formulas (5.13a)–(5.13c) are compliant with $N$ of different values of $\theta i$ on each surface of spherical particle (values of $\theta$ is between 0 and $\pi$) resulting in $3N$ of linear equations, which just can be applied to solve $3N$ of unknowns, $A_n$, $B_n$, $C_n$, and the distribution of fluid field could be obtained smoothly when $N$ is large enough.

## 5.1.3 Deduction of Particle Thermophoretic Velocity

The drag force and torque applied to the spherical particles by fluid can be determined as (Ganatos et al. 1980):

$$F = -8\pi\eta A_1 e_x \tag{5.14a}$$

$$T = -8\pi\eta C_1 e_y \tag{5.14b}$$

From the above equations, we know that only low-order coefficient $A_1$ and $C_1$ have contributed to the drag force and torque applied to the spherical particles.

Particles are freely suspended in solution, as the net force and net torque received by them are zero. Using this limitation in formulas (5.14a) and (5.14b), we can obtain:

$$A_1 = C_1 = 0 \tag{5.15}$$

By combining formula (5.15) and $3N$ linear equations generated by the formulas (5.13a)–(5.13c), the moving and rotation velocity $U$ and $\Omega$ of particles can be smoothly obtained.

## 5.2  Results and Discussions

This section will discuss the use of the boundary collocation method for solving the calculation results of thermophoresis of a single spherical particle in gaseous medium parallel to two plates.

### 5.2.1  Thermophoresis of Particle Parallel to One Single Plate

Tables 5.1 and 5.2 were compared when the spherical particle parallel to a plate (when $c \to \infty$) for the thermophoresis, in circumstances of $C_t^* = C_m^* = 0$, the results of different thermal conductivity ratio $k^*$ and separation parameters $a/b$ obtained by the boundary collocation method in two different boundary conditions of plates being compared with approximate results obtained by the reflection method, and possessing mutual verification. However, when $k^* = 0$, there is extreme situation existed in table, which is not compliant with the actual situation, main presence is made after comparison of values.

Both in Tables 5.1 and 5.2, the using of the numerical calculation of the boundary collocation method converges to present effective digits. Convergence rate is related to values of $a/b$, the lager the values are, the slower the convergence rate is, and also the more needed points to be accessed. When $a/b = 0.999$, the boundary collocation method must reach $M = 36$ and $N = 36$, or above, to be able to start convergence.

In Appendix C4, I illustrate the spherical particle with reflection method and obtain approximate analytical solution of thermophoresis in details, and the

**Table 5.1** The thermophoretic motion and rotation velocity of $C_t^* = C_m^* = 0$ when the spherical particles being parallel to a single adiabatic plate

| $a/b$ | $U/U_0$ | | $-a\Omega/U_0$ | |
|---|---|---|---|---|
| | Exact solution | Asymptotic solution | Exact solution | Asymptotic solution |
| $k^* = 0$ | | | | |
| 0.2 | 0.99953 | 0.99953 | 0.00030 | 0.00030 |
| 0.4 | 0.99684 | 0.99678 | 0.00492 | 0.00493 |
| 0.6 | 0.99172 | 0.99057 | 0.02669 | 0.02660 |
| 0.8 | 0.98853 | 0.97722 | 0.10164 | 0.09400 |
| 0.9 | 0.99789 | 0.96390 | 0.20389 | 0.16225 |
| 0.95 | 1.0223 | 0.95412 | 0.3189 | 0.21001 |
| 0.99 | 1.1450 | 0.94422 | 0.6162 | 0.25657 |
| 0.995 | 1.231 | | 0.759 | |
| 0.999 | 1.440 | | 1.063 | |
| $k^* = 10$ | | | | |
| 0.2 | 0.99828 | 0.99828 | 0.00030 | 0.00030 |
| 0.4 | 0.98640 | 0.98636 | 0.00498 | 0.00500 |
| 0.6 | 0.95251 | 0.95202 | 0.02759 | 0.02776 |
| 0.8 | 0.86845 | 0.87022 | 0.10766 | 0.10275 |
| 0.9 | 0.77441 | 0.79527 | 0.21183 | 0.18221 |
| 0.95 | 0.6864 | 0.74447 | 0.3119 | 0.23914 |
| 0.99 | 0.5324 | 0.69568 | 0.4778 | 0.29545 |
| 0.995 | 0.492 | | 0.520 | |
| 0.999 | 0.449 | | 0.559 | |

moving and rotation velocity of particles can be found in formula (C4.11a, b), the results of this calculation are also listed in the Tables 5.1 and 5.2, with the boundary collocation method to get the correct value results comparable to each other. The results showed that when $\lambda \leq 0.8$, the results of regularization moving velocity obtained by the boundary collocation method agreed well with the numerical values obtained by the reflection method, and the error is less than 1.3 %. However, when the value of $\lambda$ is larger, the accuracy of formula (C4.11a, b) is much lower.

For different thermal conductivity ratio $k^*$ and separation parameter $a/b$, when $C_t^* = C_m^* = 0.02$, the numerical results of moving and rotation velocity of normalized thermophoresis of particles $U/U_0$ and $a\Omega/U_0$ are shown in Fig. 5.2. As shown in figures, if other conditions ($C_t^*$, $C_m^*$ and $a/b$) are remained unchanged (formula of boundary conditions is (5.2c)), the regularization thermophoresis velocity in the case of the insulated plane walls, the parameter $U/U_0$ will reduce gradually with the increasing values of parameters $\kappa^*$. However, on the plates of linear temperature distribution (boundary conditions of the formula (5.2e)), then the regularization thermophoresis velocity $U/U_0$ will increase with the increasing values of parameters $\kappa^*$.

While $a/b$ increases, the mobility of particles in gaseous medium will decreases and then increases, and here is the advanced instruction: when the plate is linear

**Table 5.2** The thermophoretic motion and rotation velocity of $C_t^* = C_m^* = 0$ when the spherical particles being parallel to a single temperature

| $a/b$ | $U/U_0$ | | $-a\Omega/U_0$ | |
|---|---|---|---|---|
| | Exact solution | Asymptotic solution | Exact solution | Asymptotic solution |
| $k^* = 0$ | | | | |
| 0.2 | 0.99853 | 0.99853 | 0.00030 | 0.00030 |
| 0.4 | 0.98840 | 0.98837 | 0.00498 | 0.00500 |
| 0.6 | 0.95922 | 0.95883 | 0.02767 | 0.02774 |
| 0.8 | 0.88358 | 0.88659 | 0.10880 | 0.10261 |
| 0.9 | 0.79405 | 0.81880 | 0.21593 | 0.18187 |
| 0.95 | 0.7070 | 0.77229 | 0.3200 | 0.23865 |
| 0.99 | 0.5509 | 0.72732 | 0.4940 | 0.29480 |
| 0.995 | 0.510 | | 0.538 | |
| 0.999 | 0.465 | | 0.580 | |
| $k^* = 10$ | | | | |
| 0.2 | 0.99978 | 0.99978 | 0.00030 | 0.00030 |
| 0.4 | 0.99888 | 0.99883 | 0.00492 | 0.00493 |
| 0.6 | 0.99907 | 0.99784 | 0.02676 | 0.02658 |
| 0.8 | 1.00890 | 0.99614 | 0.10276 | 0.09386 |
| 0.9 | 1.03209 | 0.99261 | 0.20872 | 0.16192 |
| 0.95 | 1.0678 | 0.98912 | 0.3312 | 0.20952 |
| 0.99 | 1.1836 | 0.98506 | 0.6613 | 0.25592 |
| 0.995 | 1.247 | | 0.817 | |
| 0.999 | 1.391 | | 1.105 | |

temperature distributed, as the particle approaching the plate, the surrounding temperature gradient of the particle will increase dramatically with increased thermal conductivity ratio $k^*$; when the plate is insulted, as the particle approaches the plate, the surrounding temperature gradient of the particle will increase dramatically with decreased thermal conductivity ratio $k^*$, thus it will generate phenomenon of prior decrease to latter increase accompanied with increase of $a/b$ (refer to analysis in Appendix 4). When $k^* = \left(1 - C_t^*\right)^{-1}$, these two different boundary conditions of plates will have same results. The particle–plate interaction of temperature disappears in this special condition. Since there is only the effect of flow force with the presence of plates, the thermophoresis mobility of particles in gaseous status reduces monotonically with increasing $a/b$.

An observation on Tables 5.1 and 5.2 and Fig. 5.2a in detail suggests an interesting phenomenon: when the insulated plate $k^*$ is very small (i.e., $k^* = 0$), and the value of $a/b$ is smaller, the mobility of thermophoresis of particles will decrease gradually to a minimum value with increasing values of $a/b$, and will increase gradually and monotonically with constantly increasing values of $a/b$ by then. If the particle–plate gap is small enough, the velocity of its motion will be even greater than it is with the absence of plates. For example, when $C_t^* = C_m^* = 0$, $k^* = 0$ and $a/b = 0.999$, the moving velocity of the particles will be 45 % faster than that of absence of plates. In the case of higher $k^*$, the moving velocity of

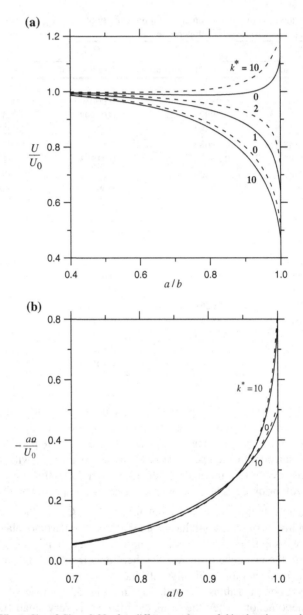

**Fig. 5.2**   **a** When $C_t^* = 2C_m^* = 0.02$, for different values of $k^*$, the thermophoretic motion velocity of $U/U_0$ versus $a/b$ of aerosol particles parallel to a single plate. **b** When $C_t^* = 2C_m^* = 0.02$, for a different values, the thermophoretic motion velocity of $a\Omega/U_0$ to $a/b$ of aerosol particles parallel to a single plate

particles parallel to an insulated plate will decrease monotonically and gradually with the increase of $a/b$. When the plate displays a linear temperature distribution, and the thermal conductance ratio $k^*$ is larger (e.g., $k^* = 10$), when $a/b$ is smaller, the thermophoresis mobility of particles will decrease to a minimum value with increasing $a/b$, and then again, increasing with incremented $a/b$. When the particle–plate gap is small enough, the motion velocity of particles will also be greater than that of absence of plates. While in the case of smaller $k^*$, the motion velocity of particles parallel to a plate with linear temperature distribution will monotonously decrease along with the increasing of $a/b$.

The interesting situation of $U/U_0$ instead of monotonic decrease is understandable: the resistance effect of the flow force is competing with acceleration of temperature gradient, which resulted in an extremely small thermal conductance ratio $k^*$ in insulated plates, and the plate with linear temperature distribution in the case of larger $k^*$, is the main factor causing the prior reduction and latter acceleration of motion velocity of the particles. And the $U/U_0$ obtained by reflection method (formula (C4.11a)) is compliant with description in Fig. 5.2a.

In Tables 5.1 and 5.2 and Fig. 5.2b, the directions of thermophoresis motion and rotation caused by body force field (such as gravity field) of spherical particle in a gaseous medium are contrary under the same geometrical conditions. The comparison is similar to electrophoresis of insulating plate parallel to the thin electric double layer of charged particles (see Keh and Chen 1988). For any thermal conductivity ratio $k^*$, under the other conditions remain unchanged (fixed $C_t^*$, $C_m^*$ and $k^*$), the rotational velocity of regularization thermophoresis of the particles is a monotonically increasing function of $a/b$, however, when $a/b$ is not large, $k^*$ is less influential for rotational velocity $a\Omega/U_0$.

In the case of $k^* = 100$, for different $C_t^* = 2C_m^*$, and the moving and rotational velocity of normalized thermophoresis of the particles, $U/U_0$ and $a\Omega/U_0$ with the separation variables $a/b$, the numerical results of them are shown in Figs. 5.3a, b and the curves therein all have its extremum. In case of heat insulated plates, the $U/U_0$ will increase along with the rising $C_t^*$, and reduce with the rising $C_m^*$. And the results of competition of these two effects had led to generation of minimum value of $U/U_0$.

On the other hand, as the plate is under boundary conditions of linear distribution, the $U/U_0$ will decrease along with the rising $C_t^*$, and increase with the rising $C_m^*$. And the results of competition of these two effects had led to generation of minimum value of $U/U_0$ (when $k^*$ is larger, its value will be far larger than 1). Generally, the extremum all occurred in the vicinity of the $C_m^* = 0.01$.

In the case of $C_t^* = C_m^* = 0.2$, different thermal conductivity ratio $k^*$, and separation of variables $a/b$ of moving velocity and rotational velocity of normalized thermophoretic of the particles, $U/U_0$ and $a\Omega/U_0$, are shown in Figs. 5.4a, b. The $U/U_0$ will appear monotonically decreasing along with the rising $k^*$ with heat insulated plate, and in boundary conditions of linear temperature, it will appear monotonically decreasing status. However, herein, the $U/U_0$ is not greater than 1 (of situation that $C_t^*$ and $C_m^*$ are larger). Similarly, $a\Omega/U_0$ with different given values of $a/b$ is less impacted by the change of $k^*$.

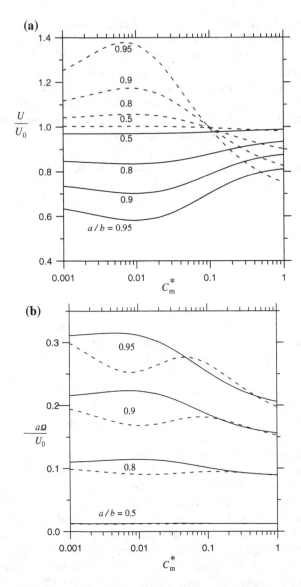

**Fig. 5.3** **a** For different separation variables $a/b$, when $k^* = 100$, the normalized thermophoretic motion velocity of $U/U_0$ to $C_m^*(C_t^* = 2C_m^*)$ of particle plotted (The *solid curves* represent the case of insulated walls, and the *dashed curves* denote the case of walls prescribed with the far-field temperature distribution). **b** For different separation variables $a/b$, when $k^* = 100$, normalized rotation thermophoretic rotation velocity of $a\Omega/U_0$ to $C_m^*(C_t^* = 2C_m^*)$ of particles (The *solid curves* represent the case of insulated walls, and the *dashed curves* denote the case of walls prescribed with the far-field temperature distribution)

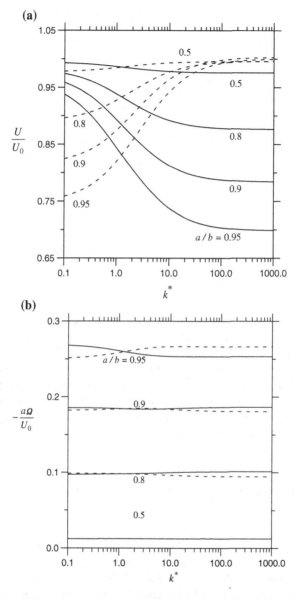

**Fig. 5.4** **a** For different separation variables $a/b$, when $C_t^* = 2C_m^* = 0.2$, normalized thermophoretic motion velocity of $U/U_0$ to $k^*$ of particles (The *solid curves* represent the case of insulated walls, and the *dashed curves* denote the case of walls prescribed with the far-field temperature distribution). **b** For different separation variables $a/b$, when $C_t^* = 2C_m^* = 0.2$, the rotation velocity of $a\Omega/U_0$ to $k^*$ of normalized thermophoretic motion of particles (The *solid curves* represent the case of insulated walls, and the *dashed curves* denote the case of walls prescribed with the far-field temperature distribution.)

**Fig. 5.5** For different values
of $C_m^*$, when $C_t = 2C_m$ and
$k^* = \left(1 - C_t^*\right)^{-1}$, the
mobility of normalized
thermophoretic motion (*solid
line*) and movable degrees
(*dotted line*) of $a/b$

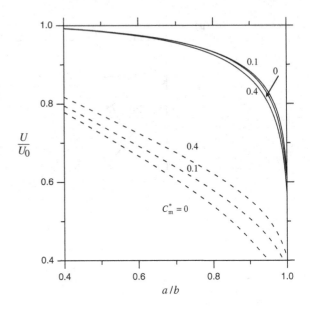

In Appendix B, using the boundary collocation method for solving a single spherical particle affected by fixed body force $Fe_x$ parallel to a large plate of effect of creeping flow. Figure 5.5 Comparing spherical particles affected by the gravitational field (in this case, $U_0 = (F/6\pi\eta a)(1 + 3C_m^*)/(1 + 2C_m^*)$) and thermophoretic force, which can be found: the affecting force of plates to particles are far less than sedimentation motion of itself. It is worth of mentioning: when $C_m$ becomes smaller, the boundary effect on particles is enhanced by the gravitational field, and of the particle the significance of thermophoresis force is weakened.

## 5.2.2 Thermophoresis of Particle Parallel to Two Plane Walls

Table 5.3 comparing spherical particles gaseous medium located between the two plates ($c = b$), possessing thermophoresis while being parallel to the two plates, for different thermal conductivity ratio $k^*$ and separation parameters $a/b$, in the situation of $C_t^* = C_m^* = 0$, with the boundary conditions of two different plates, the numerical results obtained by boundary collocation method will be mutually verified with the approximate results obtained by the reflection method (see formula (C4.20) in Appendix C).

Similar to the case of particles motion parallel to a single plate, when $\lambda \leq 0.6$ the correct results obtained by the boundary collocation method are considered similar to those obtained by reflection method; however, while $\lambda \geq 0.8$, the ones

**Table 5.3** The thermophoretic motion and rotation velocity of $C_t^* = C_m^* = 0$ and $b = c$ when the spherical particles parallel to two plane walls

| $a/b$ | $U/U_0$ | | | |
|---|---|---|---|---|
| | $k^* = 0$ | | $k^* = 10$ | |
| | Exact solution | Asymptotic solution | Exact solution | Asymptotic solution |
| *For insulated plane walls* | | | | |
| 0.2 | 0.99796 | 0.99796 | 0.99497 | 0.99497 |
| 0.4 | 0.98597 | 0.98617 | 0.96241 | 0.96288 |
| 0.6 | 0.96339 | 0.96661 | 0.88479 | 0.89411 |
| 0.8 | 0.94684 | 0.96326 | 0.74853 | 0.81948 |
| 0.9 | 0.96662 | 0.98325 | 0.64244 | 0.80784 |
| 0.95 | 1.0163 | 1.00270 | 0.5649 | 0.81679 |
| 0.99 | 1.2243 | 1.02412 | 0.4577 | 0.83402 |
| 0.995 | 1.362 | | 0.441 | |
| 0.999 | 1.712 | | 0.446 | |
| *For plane walls prescribed with the far-field temperature profile* | | | | |
| 0.2 | 0.99586 | 0.99587 | 0.99811 | 0.99811 |
| 0.4 | 0.96931 | 0.96975 | 0.98717 | 0.98736 |
| 0.6 | 0.90602 | 0.91448 | 0.96745 | 0.97058 |
| 0.8 | 0.79069 | 0.85485 | 0.95706 | 0.97239 |
| 0.9 | 0.69390 | 0.84471 | 0.98248 | 0.99596 |
| 0.95 | 0.6184 | 0.85078 | 1.0352 | 1.01744 |
| 0.99 | 0.5075 | 0.86316 | 1.2164 | 1.04060 |
| 0.995 | 0.490 | | 1.324 | |
| 0.999 | 0.495 | | 1.582 | |

obtained by reflection method exists considerablly error. Broadly speaking, the formula (C4.20) overestimated thermophoretic velocity of the particle.

Comparing Table 5.3 with Tables 5.1 and 5.2, we can find out: when $a/b$ is smaller, directly summing of boundary effects of a single plate will underestimate boundary effects of two plates; However, when $a/b$ is larger, the same behavior will overestimate boundary effects of two plates.

For the moving velocity $U/U_0$ of normalized thermophoresis of particles with different $C_t^*$, $C_m^*$, $k^*$ and separation parameter $a/b$, the numerical results obtained by the boundary collocation method are shown in Figs. 5.6a, b. As shown in figures, the velocity $U/U_0$ of normalized thermophoresis will decrease along with the increasing $k^*$ with heat insulated plates; and the velocity $U/U_0$ of normalized thermophoresis will increase along with the increasing $k^*$ with plates of linear temperature distribution. Similarly, with heat insulated plates and $k^*$ is extremely small, when a/b is smaller, the mobility of thermophoresis of particles will in decrease along with the increasing $a/b$ to its minimum value, and then, increase with the increasing $a/b$. While particle–plate gap is extremely small, the motion velocity of particles will also be greater than the status with absence of plates. Therefore, when the particle–plate gap is small enough, which will make greater acceleration force of temperature gradient than resistance force of flow and the

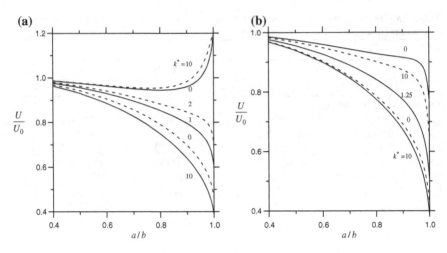

**Fig. 5.6** **a** When $C_t^* = 2C_m^* = 0.02$ and $b = c$, the thermophoretic motion velocity of $U/U_0$ to $a/b$ of aerosol particles parallel to the two plane walls (The *solid curves* represent the case of insulated walls, and the *dashed curves* denote the case of walls prescribed with the far-field temperature distribution.). **b** When $C_t^* = 2C_m^* = 0.2$ and $b = c$, the thermophoretic motion velocity of $U/U_0$ to $a/b$ of aerosol particles parallel to the two plane walls (The *solid curves* represent the case of insulated walls, and the *dashed curves* denote the case of walls prescribed with the far-field temperature distribution.)

motion velocity will be accelerated since that. As for the approximate results obtained by reflection method (formula (C4.20)), its trend is compliant with that of boundary collocation method.

The comparison between Figs. 5.2a and 5.6a shows that, when adding the second plate, it may not enhance its influence on velocity thermophoresis of the particles (even with the equal distance from the particles to both plates). When adding a second plate, although the resistance of the current force and the growth rate of the temperature gradient are all enhanced, the levels of enhancement are varying, and thus the total influence may not be able to increase the velocity of thermophoretic of the particle. The particle in a gaseous medium located in an arbitrary position between the two plates, for different separation parameters $a/b$, when $k^* = \left(1 - C_t^*\right)^{-1}$ and the $C_t^* = 2C_m^* = 0.2$ (in this case of the boundary conditions of two different plates have the same numerical results), the moving and rotational velocity of normalized thermophoresis of the particle, $U/U_0$ and $a\Omega/U_0$, are shown in Figs. 5.7a, b. The dotted line represents the fixed distance between a plate and particle ($a/b = $ constant), the influence on thermophoresis of particles caused by changes in the other plate (at $z = c$). The solid line represents the pitch between two plates is fixed ($2a/(b + c) = $ constant), the influence on thermophoresis of particles is caused by particle location in different positions between the two plates. As shown in Fig. 5.7a, we can find out that under a given situation, the net effect of plates will decelerate velocity $U/U_0$ of thermophoresis of particles. When $2a/(b + c)$ is fixed, as particle location at the middle of the two plates (when

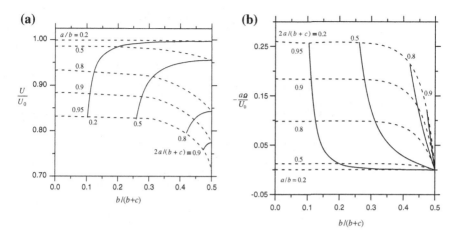

**Fig. 5.7** **a** When $C_t^* = 2C_m^* = 0.2$ and $k^* = \left(1 - C_t^*\right)^{-1}$, the thermophoretic motion velocity of $U/U_0$ to $b/(b + c)$ of aerosol particles parallel to the two plane walls. **b** When $C_t^* = 2C_m^* = 0.2$ and $k^* = \left(1 - C_t^*\right)^{-1}$, the thermophoretic motion velocity of $a\Omega/U_0$ to $b/(b + c)$ of aerosol particles parallel to the two plane walls

$c = b$), there is existing the smallest viscous drag force leading to maximum moving velocity (the rotational velocity is zero at the time). While particles approaching palates ($b/(b + c)$ becoming small), the drag force of fluid will increase leading to deceleration of moving velocity, but the rotational velocity is accelerated.

While fixing the distance between particles and one of the plates (when $a/b$ is fixed), the presence of the other plate will reduce moving and rotational velocity of particles; and along with approaching of particles to the other plate (when $b/(b + c)$ becomes large), the moving and rotational velocity of particles will decrease gradually.

In other words, when the two plates are heat insulated the thermal conductivity ratio $k^*$ is minimal or two plates are a linear temperature distribution, in case of case of $k^*$ is larger, as shown in Figs. 5.8a, b, the net effect of plates on the particles will accelerate the velocity of its thermophoresis.

In Figs. 5.8a, b, it is described when $k^* = 100$, for the different $C_t^* \left(= 2C_m^*\right)$ and the separation variables $a/b$, the moving and rotational velocity of normalized thermophoresis of particles, respectively, $U/U_0$ and $a\Omega/U_0$, can be compared to each other with Figs. 5.7a, b.

Cross-reference to several different $2a/(b + c)$ case suggests that when the particle is located at the middle of the two plates, there is relative maximum value of the particle velocity; when particles near any one of the plates, its relative velocity will be decreased gradually. It is worth mentioning that the fixed $a/b$ value in the fixed particles and wherein the distance of a plate, the other plate, the effect of which on particle velocity will not be monotonic function, however, for the sake of brevity, this is no longer plotted instructions.

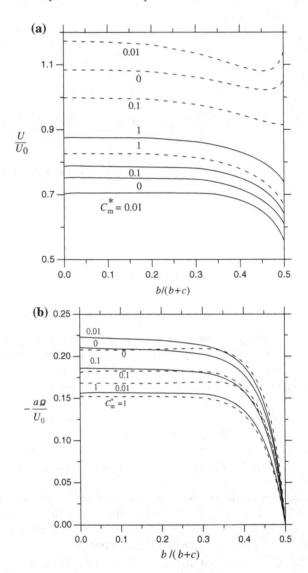

**Fig. 5.8** **a** When $C_t^* = 2C_m^*$, $k^* = 100$ and $a/b = 0.9$, the thermophoretic motion velocity of $U/U_0$ to $b/(b+c)$ of aerosol particles parallel to the two plane walls (The *solid curves* represent the case of insulated walls, and the *dashed curves* denote the case of walls prescribed with the far-field temperature distribution.). **b** When $C_t^* = 2C_m^*$, $k^* = 100$ and $a/b = 0.9$, the thermophoretic motion velocity of $a\Omega/U_0$ to $b/(b+c)$ of aerosol particles parallel to the two plane walls (The *solid curves* represent the case of insulated walls, and the *dashed curves* denote the case of walls prescribed with the far-field temperature distribution.)

In Appendix B, the boundary collocation method is used for solving a single spherical particle in the slip case, and the particle is located in an arbitrary position between the two plates, which is sedimentation motion in parallel to the two plates. Comparing with this chapter, the plate effects of thermophoresis are even less than that of sedimentation.

## 5.3 Conclusions

This study considers single spherical colloidal particles in the case of the low Reynolds number and low pictogram number of columns, parallel in a single infinite plate or infinite plate of thermophoretic motion behavior. Respectively, taking point method (boundary collocation method) with the reflection method for solving particle swimming velocity and compare particle phoretic motion in the case.

In this chapter, I calculate a single spherical particle in a gaseous medium at a fixed temperature gradient effecting velocity of thermocapillary motion. The boundary conditions of the plate should be heat insulated with linear temperature distribution plate kinds of situations are discussed. Overall, the thermophoretic velocity purposes, and also showed a monotonically decreasing situation. When $a/b$ tends to 1, for the different values of $k^*$, the different plate boundary condition increase or decrease the particle thermophoretic velocity. This phenomenon is also a temperature gradient enhancement effect and fluid effect competing with the results.

## References

Brock, J.R.: On the theory of thermal forces acting on aerosol particles. J. Colloid Sci. **17**, 768 (1962)

Ganatos, P., Weinbaum, S., Pfeffer, R.: A strong interaction theory for the creeping motion of a sphere between plane parallel boundaries. Part 2. Parallel motion. J. Fluid Mech. **99**, 755 (1980)

Keh, H. J., Chen, S. B.: Electrophoresis of a colloidal sphere parallel to a dielectricplane. J. Fluid Mech. **194**, 377 (1988)

# Chapter 6
# General Discussions and Conclusions

**Abstract** In this chapter, the above four movement actions will be compared in terms of the similarities and differences, and analyzed in the tables, so as to understand various motion phenomena.

## 6.1 General Discussions

In various phoretic motion stated in the previous chapters, we can find much similarity in different phoretic motions, but there are also some differences. These are summarized in Table 6.1. Various phoretic motions are generated by external field $\nabla Y\infty$, which may be a temperature gradient or concentration gradient, etc. In this study, we only discuss the case that $\nabla Y\infty$ is constant, therefore the motion velocity of colloidal particles (osmophoretic motion with vesicle particle; thermocapillary motion with droplets) is proportional to the external field $\nabla Y\infty$. Diffusiophoretic or osmophoretic motions are fluid with external solute concentration gradient causing phoretic motions of colloidal particles or vesicle particles. Thermocapillary or thermophoretic motions are generated by the imposed external temperature gradient in the fluid causing the phoretic motions of spherical droplets or particles in gaseous medium.

The variables in diffusiophoresis are $L*$, the featured length of the interaction between the particles and solute molecules; $K$, is solute molecules adsorbed on the particle surface related to the degree of Gibbs adsorption length; $\beta$ is the coefficient for the polarization; $a$ is the radius of the particles; $\eta$ is the viscosity of the fluid; $k$ is the Boltzmann constant; $T$ is the absolute temperature; and the major impact parameter is $\beta/a$; all of the above are listed in the first row of Table 6.1.

The osmophoresis related variables are $Lp$ is hydraulic coefficient; $a$ is radius of vesicle particles; $R$ is the ideal gas constant; $T$ is the absolute temperature; and the major impact parameters are $\kappa$ and $\bar{\kappa}$; all of the above are listed in the second row of Table 6.1.

P.-Y. Chen, *The Application of Biofluid Mechanics*, SpringerBriefs in Physics,         87
DOI: 10.1007/978-3-642-44952-9_6, © The Author(s) 2014

**Table 6.1** Similarity among the four phoretic motions as well as the associated parameters with various phoretic motions

| | $U = \alpha\nabla Y_\infty$ | | |
|---|---|---|---|
| | Field variable ($Y_\infty$) | $\alpha$ | Parameters |
| Diffusiophoresis | Concentration | $\frac{kT}{\eta} L * K(1 + \frac{\beta}{a})^{-1}$ | $\beta/a$ |
| Osmophoresis | Concentration | $-aL_p RT (2 + 2\bar{\kappa} + \kappa)^{-1}$ $\kappa = aL_p RTC_0/D,\ \bar{\kappa} = aL_p RT\bar{C}/\bar{D}$ | $\kappa, \bar{\kappa}$ |
| Thermocapillary motion | Temperature | $\frac{2}{(2+k^*)(2+3\eta^*)}\left(-\frac{\partial\gamma}{\partial T}\right)\frac{a}{\eta}$ $k^* = k_1/k,\ \eta^* = \eta_1/\eta$ | $k^*, \eta^*$ |
| Thermophoresis | Temperature | $\left[\frac{2C_s(k+k_1 C_t^*)}{(1+2C_m^*)(2k+k_1+2kC_t^*)}\right]\frac{\eta}{\rho_f T_0}$ $k^* = k_1/k,\ C_t^* = C_t l/a,\ C_m^* = C_m l/a$ | $k^*, C_t^*, C_m^*$ |

Thermocapillary motion related variables are: $k^*$ is the ratio of the thermal conductivity ratio of the fluid inside and outside the droplets; $\eta^*$ is the ratio of viscosity of the fluid inside and outside of the droplets; $a$ is radius of a spherical droplet; $\eta$, is the viscosity of the fluid; $\partial\gamma/\partial T$ is a surface tension $\gamma$ with the changes of the local temperature gradient $T$; and the major impact parameter $k^*$ and $\eta^*$; all of the above are listed the third row of Table 6.1.

Thermophoresis related variables are: $\rho_f$ is the density of the gas; $\eta$ is the viscosity of the fluid; $T_0$ is the absolute temperature of particles center when the particles do not exist; $l/a$ is the Knudsen number; $k$ the thermal conductivity of gas; $k_1$ is the thermal conductivity of the particles; $Cs$ is the thermal slip coefficient; $C_t^{**}$ is the temperature jump differential coefficient multiplied by $l/a$; $C_m^*$ is the friction slip coefficient multiplied by $l/a$; and the major impact parameters are $k^*$, $C_t^*$ and $C_m^*$; all of the above are listed in the fourth row of Table 6.1.

Table 6.1 shows the similarity among these four phoretic motions as well as the associated parameters with various phoretic motions. In this study, using the consistent coordinates for every phoretic motion in parallel to plates, the direction of phoretic motion of particles is the positive x-axis direction, and the direction of the normal direction of plates for the z-axis direction, to simplify the analog of four phoretic motions. Table 6.2 shows the comparison of the concentration gradient as the driving force of diffusiophoresis with osmophoresis sorted out by the important results of the reflection method in Appendix C to its motion characteristics. Among them, the first row is the phoretic motion velocity results related to Faxen's Law derived. The second row is the relevant parameters $G$ closely related to the effect of concentration fields. The third row indicates that when the parameter $G$ is equal to 0, under special circumstances, the plates of two different boundary conditions (solute impermeable as well as the solute into the linear concentration distribution) with calculation of the same phoretic motion velocity; diffusiophoresis is when $\beta/a = 1/2$, osmophoresis is when $1+\bar{\kappa} = \kappa$. The fourth row

**Table 6.2** Comparison of the concentration gradient as the driving force of diffusiophoresis with osmophorsis

|  | Diffusiophorsis | Osmophorsis |
|---|---|---|
| Faxen's law | $\mathbf{U}^{(i)} = A[\nabla C_w^{(i)}]_0 + [v_w^{(i)}]_0 + \frac{a^2}{6}[\nabla^2 v_w^{(i)}]_0$ | $\mathbf{U}^{(i)} = -A[\nabla C_w^{(i)}]_0 + [v_w^{(i)}]_0 + \frac{a^2}{6}[\nabla^2 \mathbf{v}_w^{(i)}]_0$ |
|  | $\mathbf{\Omega}^{(i)} = \frac{1}{2}[\nabla \times \mathbf{v}_w^{(i)}]_0$ | $\mathbf{\Omega}^{(i)} = \frac{1}{2}[\nabla \times \mathbf{v}_w^{(i)}]_0$ |
|  | $A = \frac{kT}{\eta}L^*K(1+\frac{\beta}{a})^{-1}$ | $A = aL_pRT(2+2\bar{\kappa}+\kappa)^{-1}$ |
| Parameter $G$ | $(1/2 - \beta/a)(1+\beta/a)^{-1}$ | $(1+\bar{\kappa}-\kappa)(2+2\bar{\kappa}+\kappa)^{-1}$ |
| Special case of $G = 0$ | $\beta/a = 1/2$ | $1+\bar{\kappa} = \kappa$ |
| Direction of $\nabla C_\infty$ | $+\mathbf{e}_x$ | $-\mathbf{e}_x$ |
| Direction of $\mathbf{\Omega}$ | $-\mathbf{e}_y$ | $+\mathbf{e}_y$ |

**Table 6.3** Comparison of thermocapillary motion and thermophoresis driven by the temperature gradient

|  | Thermocapillary motion | Thermophoresis |
|---|---|---|
| Faxen's law | $\mathbf{U}^{(i)} = A\frac{a}{\eta}(-\frac{\partial \gamma}{\partial T})[\nabla T_w^{(i)}]_0 + [v_w^{(i)}]_0$ | $\mathbf{U}^{(i)} = A[\nabla T_w^{(i)}]_0 + [v_w^{(i)}]_0 + \frac{a^2 D}{6}[\nabla^2 v_w^{(i)}]_0$ |
|  | $+\frac{a^2 C}{6}[\nabla^2 \mathbf{v}_w^{(i)}]_0$ | $\mathbf{\Omega}^{(i)} = \frac{1}{2}[\nabla \times \mathbf{v}_w^{(i)}]_0$ |
|  | $A = 2(2+k^*)^{-1}(2+3\eta^*)^{-1}$ | $A = \frac{2C_s(k+k_1C_t l/a)}{(1+2C_m l/a)(2k+k_1+2kC_t l/a)}\frac{\eta}{\rho_f T_0}$ |
| Parameter $G$ | $(1-k^*)(2+k^*)^{-1}$ | $(1-k^*+k^*C_t^*)(2+k^*+2k\cdot C_t^*)^{-1}$ |
| Special case of $G = 0$ | $k^* = 1$ | $k^* = (1-C_t^*)^{-1}$ |
| Direction of $\nabla T_\infty$ | $+\mathbf{e}_x$ | $-\mathbf{e}_x$ |
| Direction of $\mathbf{\Omega}$ | $\times$ | $-\mathbf{e}_y$ |

represents the direction of the solute concentration gradient, due to the diffusiophoresis of colloidal particles toward the direction of high solute concentration due to diffusion, the gradient direction is the same direction of the x-axis coordinates, and the osmophoresis of vesicle particles move toward the direction of the low solute concentration, therefore gradient direction opposite to the direction of the x-axis coordinates. The fifth row is the rotation angular velocity direction particle.

Table 6.3 shows the comparison of thermocapillary motion and thermophoresis driven by the temperature gradient and is sorted out by the important results of the reflection method in Appendix C of its motion characteristics. Among them, the first row is the phoretic motion velocity results derived from Faxen's Law. The second row is the relevant parameter $G$ closely related to effect of temperature field. The third row is when the parameter $G$ is equal to 0, under the special situation, the plates of two different boundary conditions (heat insulation and the linear temperature distribution) with the calculation of the same phoretic motion velocity; thermocapillary motion is when $k^* = 1$, while the thermophoretsis is when $k^* = \left(1 - C_t^*\right)^{-1}$. The fourth row is the direction of the temperature gradient, due to the thermocapillary motion is the droplets moving toward the direction of high temperature, the gradient direction is the same as the x-axis coordinates, while thermophoresis of particles in gaseous medium moving toward the direction of low temperature, therefore the gradient direction is opposite to x-direction. The fifth row is the angular velocity of rotational direction of the particle, due to the particles of thermocapillary motion is droplets, and there is no rotating available.

## 6.2 Conclusions

In this study, I explore a single spherical colloidal particles (consider the vesicle particle semipermeable membrane; the thermocapillary then should consider droplets) parallel single infinite plate or infinite plate phoretic motion boundary effects, and compare the similarities and differences between the various phoretic motions. It is hoped that this research and application can provide a base for research theory analysis and will be helpful.